T0260077

Algebraic Geometry in Coding Theory and Cryptography

Algebraic Geometry in Coding Theory and Cryptography

HARALD NIEDERREITER
AND CHAOPING XING

PRINCETON UNIVERSITY PRESS
PRINCETON AND OXFORD

Library of Congress Cataloging-in-Publication Data

Niederreiter, Harald, 1944–
Algebraic geometry in coding theory and cryptography / Harald
Niederreiter and Chaoping Xing.
p.   cm.
Includes bibliographical references and index.
ISBN 978-0-691-10288-7 (hardcover : alk. paper)
1. Coding theory.   2. Cryptography.   3. Geometry, Algebraic.
I. Xing, Chaoping, 1963–   II. Title.
QA268.N54 2009
003'. 54—dc22   2008056156

British Library Cataloging-in-Publication Data is available

This book has been composed in Times
Printed on acid-free paper. ∞
press.princeton.edu

Typeset by S R Nova Pvt Ltd, Bangalore, India
Printed in the United States of America

10 9 8 7 6 5 4 3 2 1

For Gerlinde, spirited and indomitable companion on that great trek called life

To my wife Youqun Shi and my children Zhengrong and Menghong

Contents

Preface

Algebraic geometry has found fascinating applications to coding theory and cryptography in the last few decades. This book aims to provide the necessary theoretical background for reading the contemporary literature on these applications. An aspect that we emphasize, as it is very useful for the applications, is the interplay between nonsingular projective curves over finite fields and global function fields. This correspondence is well known and frequently employed by researchers, but nevertheless it is difficult to find detailed proofs of the basic facts about this correspondence in the expository literature. One contribution of our book is to fill this gap by giving complete proofs of these results.

We also want to offer the reader a taste of the applications of algebraic geometry, and in particular of algebraic curves over finite fields, to coding theory and cryptography. Several books, among them our earlier book *Rational Points on Curves over Finite Fields: Theory and Applications*, have already treated such applications. Accordingly, besides presenting standard topics such as classical algebraic-geometry codes, we have also selected material that cannot be found in other books, partly because it is of recent origin.

As a reflection of the above aims, the book splits into two parts. The first part, consisting of Chapters 1 to 4, develops the theory of algebraic varieties, of algebraic curves, and of their function fields, with the emphasis gradually shifting to global function fields. The second part consists of Chapters 5 and 6 and describes applications to coding theory and cryptography, respectively. The book is written at the level of advanced undergraduates and first-year graduate students with a good background in algebra.

We are grateful to our former Ph.D. students David Mayor and Ayineedi Venkateswarlu for their help with typesetting and proofreading. We also thank Princeton University Press for the invitation to write this book.

Singapore, November 2007   HARALD NIEDERREITER
                                              CHAOPING XING

Algebraic Geometry in Coding
Theory and Cryptography

1   Finite Fields and Function Fields

In the first part of this chapter, we describe the basic results on finite fields, which are our ground fields in the later chapters on applications. The second part is devoted to the study of function fields.

Section 1.1 presents some fundamental results on finite fields, such as the existence and uniqueness of finite fields and the fact that the multiplicative group of a finite field is cyclic. The algebraic closure of a finite field and its Galois group are discussed in Section 1.2. In Section 1.3, we study conjugates of an element and roots of irreducible polynomials and determine the number of monic irreducible polynomials of given degree over a finite field. In Section 1.4, we consider traces and norms relative to finite extensions of finite fields.

A function field governs the abstract algebraic aspects of an algebraic curve. Before proceeding to the geometric aspects of algebraic curves in the next chapters, we present the basic facts on function fields. In particular, we concentrate on algebraic function fields of one variable and their extensions including constant field extensions. This material is covered in Sections 1.5, 1.6, and 1.7.

One of the features in this chapter is that we treat finite fields using the Galois action. This is essential because the Galois action plays a key role in the study of algebraic curves over finite fields. For comprehensive treatments of finite fields, we refer to the books by Lidl and Niederreiter [71, 72].

1.1 Structure of Finite Fields

For a prime number p, the residue class ring $\mathbb{Z}/p\mathbb{Z}$ of the ring $\mathbb{Z}$ of integers forms a field. We also denote $\mathbb{Z}/p\mathbb{Z}$ by $\mathbb{F}_p$. It is a prime field in the sense that there are no proper subfields of $\mathbb{F}_p$. There are exactly p elements in $\mathbb{F}_p$. In general, a field is called a *finite field* if it contains only a finite number of elements.

Proposition 1.1.1. Let k be a finite field with q elements. Then:

(i) there exists a prime p such that $\mathbb{F}_p \subseteq k$;
(ii) $q = p^n$ for some integer $n \geq 1$;
(iii) $\alpha^q = \alpha$ for all $\alpha \in k$.

Proof.

(i) Since k has only $q < \infty$ elements, the characteristic of k must be a prime p. Thus, $\mathbb{F}_p$ is the prime subfield of k.

(ii) We consider k as a vector space over $\mathbb{F}_p$. Since k is finite, the dimension $n := \dim_{\mathbb{F}_p}(k)$ is also finite. Let $\{\alpha_1, \ldots, \alpha_n\}$ be a basis of k over $\mathbb{F}_p$. Then each element of k can be uniquely represented in the form $a_1\alpha_1 + \cdots + a_n\alpha_n$ with $a_1, \ldots, a_n \in \mathbb{F}_p$. Thus, $q = p^n$.

(iii) It is trivial that $\alpha^q = \alpha$ if $\alpha = 0$. Assume that α is a nonzero element of k. Since all nonzero elements of k form a multiplicative group k^* of order $q - 1$, we have $\alpha^{q-1} = 1$, and so $\alpha^q = \alpha$. $\qquad\square$

Using the above proposition, we can show the most fundamental result concerning the existence and uniqueness of finite fields.

Theorem 1.1.2. For every prime p and every integer $n \geq 1$, there exists a finite field with p^n elements. Any finite field with $q = p^n$ elements is isomorphic to the splitting field of the polynomial $x^q - x$ over $\mathbb{F}_p$.

Proof. (Existence) Let $\overline{\mathbb{F}_p}$ be an algebraic closure of $\mathbb{F}_p$ and let $k \subseteq \overline{\mathbb{F}_p}$ be the splitting field of the polynomial $x^{p^n} - x$ over $\mathbb{F}_p$. Let R be the set of all roots of $x^{p^n} - x$ in k. Then R has exactly p^n elements since the derivative of the polynomial $x^{p^n} - x$ is $p^n x^{p^n-1} - 1 = -1$. It is easy to verify that R contains $\mathbb{F}_p$ and R forms a subfield of $\overline{\mathbb{F}_p}$ (note that $(\alpha + \beta)^{p^m} = \alpha^{p^m} + \beta^{p^m}$ for any $\alpha, \beta \in \overline{\mathbb{F}_p}$ and any integer $m \geq 1$). Thus, R is exactly the splitting field k, that is, k is a finite field with p^n elements.

(Uniqueness) Let $k \subseteq \overline{\mathbb{F}_p}$ be a finite field with q elements. By Proposition 1.1.1(iii), all elements of k are roots of the polynomial $x^q - x$. Thus, k is the splitting field of the polynomial of $x^q - x$ over $\mathbb{F}_p$. This proves the uniqueness. $\qquad\square$

The above theorem shows that for given $q = p^n$, the finite field with q elements is unique in a fixed algebraic closure $\overline{\mathbb{F}_p}$. We denote this finite field by $\mathbb{F}_q$ and call it *the* finite field of order q (or with q elements). It follows from the proof of the above theorem that $\mathbb{F}_q$ is the splitting field of the polynomial $x^q - x$ over $\mathbb{F}_p$, and so $\mathbb{F}_q/\mathbb{F}_p$ is a Galois extension of degree n. The following result yields the structure of the Galois group $\mathrm{Gal}(\mathbb{F}_q/\mathbb{F}_p)$.

Lemma 1.1.3. The Galois group $\mathrm{Gal}(\mathbb{F}_q/\mathbb{F}_p)$ with $q = p^n$ is a cyclic group of order n with generator $\sigma : \alpha \mapsto \alpha^p$.

Proof. It is clear that σ is an automorphism in $\mathrm{Gal}(\mathbb{F}_q/\mathbb{F}_p)$. Suppose that σ^m is the identity for some $m \geq 1$. Then $\sigma^m(\alpha) = \alpha$, that is, $\alpha^{p^m} - \alpha = 0$, for all $\alpha \in \mathbb{F}_q$. Thus, $x^{p^m} - x$ has at least $q = p^n$ roots. Therefore, $p^m \geq p^n$, that is, $m \geq n$. Hence, the order of σ is equal to n since $|\mathrm{Gal}(\mathbb{F}_q/\mathbb{F}_p)| = n$.   □

Lemma 1.1.4. The field $\mathbb{F}_{p^m}$ is a subfield of $\mathbb{F}_{p^n}$ if and only if m divides n.

Proof. If m divides n, then there exists a subgroup H of $\mathrm{Gal}(\mathbb{F}_{p^n}/\mathbb{F}_p)$ with $|H| = n/m$ since $\mathrm{Gal}(\mathbb{F}_{p^n}/\mathbb{F}_p)$ is a cyclic group of order n by Lemma 1.1.3. Let k be the subfield of $\mathbb{F}_{p^n}/\mathbb{F}_p$ fixed by H. Then $[k : \mathbb{F}_p] = m$. Thus, $k = \mathbb{F}_{p^m}$ by the uniqueness of finite fields.

Conversely, let $\mathbb{F}_{p^m}$ be a subfield of $\mathbb{F}_{p^n}$. Then the degree $m = [\mathbb{F}_{p^m} : \mathbb{F}_p]$ divides the degree $n = [\mathbb{F}_{p^n} : \mathbb{F}_p]$.   □

Theorem 1.1.5. Let q be a prime power. Then:

   (i) $\mathbb{F}_q$ is a subfield of $\mathbb{F}_{q^n}$ for every integer $n \geq 1$.
  (ii) $\mathrm{Gal}(\mathbb{F}_{q^n}/\mathbb{F}_q)$ is a cyclic group of order n with generator
     $\sigma : \alpha \mapsto \alpha^q$.
 (iii) $\mathbb{F}_{q^m}$ is a subfield of $\mathbb{F}_{q^n}$ if and only if m divides n.

Proof.

   (i) Let $q = p^s$ for some prime p and integer $s \geq 1$. Then by
     Lemma 1.1.4, $\mathbb{F}_q = \mathbb{F}_{p^s} \subseteq \mathbb{F}_{p^{ns}} = \mathbb{F}_{q^n}$.
  (ii) Using exactly the same arguments as in the proof of Lemma 1.1.3
     but replacing p by q, we obtain the proof of (ii).
 (iii) By Lemma 1.1.4, $\mathbb{F}_{q^m} = \mathbb{F}_{p^{ms}}$ is a subfield of $\mathbb{F}_{q^n} = \mathbb{F}_{p^{ns}}$ if and
     only if ms divides ns. This is equivalent to m dividing n.   □

We end this section by determining the structure of the multiplicative group $\mathbb{F}_q^*$ of nonzero elements of a finite field $\mathbb{F}_q$.

Proposition 1.1.6. The multiplicative group $\mathbb{F}_q^*$ is cyclic.

Proof. Let $t \leq q - 1$ be the largest order of an element of the group $\mathbb{F}_q^*$. By the structure theorem for finite abelian groups, the order of any element of $\mathbb{F}_q^*$ divides t. It follows that every element of $\mathbb{F}_q^*$ is a root of the polynomial $x^t - 1$, hence, $t \geq q - 1$, and so $t = q - 1$.   □

Definition 1.1.7. A generator of the cyclic group $\mathbb{F}_q^*$ is called a *primitive element* of $\mathbb{F}_q$.

Let γ be a generator of $\mathbb{F}_q^*$. Then γ^n is also a generator of $\mathbb{F}_q^*$ if and only if $\gcd(n, q - 1) = 1$. Thus, we have the following result.

Corollary 1.1.8. There are exactly $\phi(q - 1)$ primitive elements of $\mathbb{F}_q$, where ϕ is the Euler totient function.

1.2 Algebraic Closure of Finite Fields

Let p be the characteristic of $\mathbb{F}_q$. It is clear that the algebraic closure $\overline{\mathbb{F}_q}$ of $\mathbb{F}_q$ is the same as $\overline{\mathbb{F}_p}$.

Theorem 1.2.1. The algebraic closure of $\mathbb{F}_q$ is the union $\cup_{n=1}^{\infty} \mathbb{F}_{q^n}$.

Proof. Put $U := \cup_{n=1}^{\infty} \mathbb{F}_{q^n}$. It is clear that U is a subset of $\overline{\mathbb{F}_q}$ since $\mathbb{F}_{q^n}$ is a subset of $\overline{\mathbb{F}_p}$. It is also easy to verify that U forms a field.

Let $f(x) = \sum_{i=0}^{s} \lambda_i x^i$ be a nonconstant polynomial over U. Then for $0 \leq i \leq s$ we have $\lambda_i \in \mathbb{F}_{q^{m_i}}$ for some $m_i \geq 1$. Hence, by Theorem 1.1.5(iii), $f(x)$ is a polynomial over $\mathbb{F}_{q^m}$, where $m = \prod_{i=0}^{s} m_i$. Let α be a root of $f(x)$. Then $\mathbb{F}_{q^m}(\alpha)$ is an algebraic extension of $\mathbb{F}_{q^m}$ and $\mathbb{F}_{q^m}(\alpha)$ is a finite-dimensional vector space over $\mathbb{F}_{q^m}$. Hence, $\mathbb{F}_{q^m}(\alpha)$ is also a finite field containing $\mathbb{F}_q$. Let r be the degree of $\mathbb{F}_{q^m}(\alpha)$ over $\mathbb{F}_{q^m}$. Then $\mathbb{F}_{q^m}(\alpha)$ contains exactly q^{rm} elements, that is, $\mathbb{F}_{q^m}(\alpha) = \mathbb{F}_{q^{rm}}$. So α is an element of U. This shows that U is the algebraic closure $\overline{\mathbb{F}_q}$.   □

We are going to devote the rest of this section to the study of the Galois group $\mathrm{Gal}(\overline{\mathbb{F}_q}/\mathbb{F}_q)$. We start from the definition of the inverse limit for finite groups. For a detailed discussion of inverse limits of groups, we refer to the book by Wilson [130].

A *directed set* is a nonempty partially ordered set I such that for all $i_1, i_2 \in I$, there is an element $j \in I$ for which $i_1 \leq j$ and $i_2 \leq j$.

Definition 1.2.2. An *inverse system* $\{G_i, \varphi_{ij}\}$ of finite groups indexed by a directed set I consists of a family $\{G_i : i \in I\}$ of finite groups and a family $\{\varphi_{ij} \in \text{Hom}(G_j, G_i) : i, j \in I, i \leq j\}$ of maps such that φ_{ii} is the identity on G_i for each i and $\varphi_{ij} \circ \varphi_{jk} = \varphi_{ik}$ whenever $i \leq j \leq k$. Here, $\text{Hom}(G_j, G_i)$ denotes the set of group homomorphisms from G_j to G_i.

For an inverse system $\{G_i, \varphi_{ij}\}$ of finite groups indexed by a directed set I, we form the Cartesian product $\prod_{i \in I} G_i$, viewed as a product group. We consider the subset of $\prod_{i \in I} G_i$ given by

$$D := \left\{ (x_i) \in \prod_{i \in I} G_i : \varphi_{ij}(x_j) = x_i \quad \text{for all } i, j \in I \quad \text{with } i \leq j \right\}.$$

It is easy to check that D forms a subgroup of $\prod_{i \in I} G_i$. We call D the *inverse limit* of $\{G_i, \varphi_{ij}\}$, denoted by $\varprojlim G_i$.

Example 1.2.3. Define a partial order in the set $\mathbb{N}$ of positive integers as follows: for $m, n \in \mathbb{N}$, let $m \preceq n$ if and only if m divides n. For each positive integer i, let G_i be the cyclic group $\mathbb{Z}/i\mathbb{Z}$, and for each pair $(i, j) \in \mathbb{N}^2$ with $i|j$, define $\varphi_{ij} : \bar{n} \in G_j \mapsto \bar{n} \in G_i$, with the bar indicating the formation of a residue class. Then it is easy to verify that the family $\{\mathbb{Z}/i\mathbb{Z}, \varphi_{ij}\}$ forms an inverse system of finite groups indexed by $\mathbb{N}$. The inverse limit $\varprojlim \mathbb{Z}/i\mathbb{Z}$ is denoted by $\hat{\mathbb{Z}}$.

Example 1.2.4. Now let $\mathbb{F}_q$ be the finite field with q elements. We consider the family of Galois groups $G_i := \text{Gal}(\mathbb{F}_{q^i}/\mathbb{F}_q)$ of $\mathbb{F}_{q^i}$ over $\mathbb{F}_q$ for each $i \in \mathbb{N}$. We define a partial order in $\mathbb{N}$ as in Example 1.2.3. For each pair $(i, j) \in \mathbb{N}^2$ with $i|j$, define the homomorphism $\varphi_{ij} : \sigma_j \in \text{Gal}(\mathbb{F}_{q^j}/\mathbb{F}_q) \mapsto \sigma_j|_{\mathbb{F}_{q^i}} \in \text{Gal}(\mathbb{F}_{q^i}/\mathbb{F}_q)$, where $\sigma_j|_{\mathbb{F}_{q^i}}$ stands for the restriction of σ_j to $\mathbb{F}_{q^i}$. Then $\{\text{Gal}(\mathbb{F}_{q^i}/\mathbb{F}_q), \varphi_{ij}\}$ forms an inverse system of finite groups indexed by $\mathbb{N}$.

Theorem 1.2.5. We have

$$\text{Gal}(\overline{\mathbb{F}_q}/\mathbb{F}_q) \simeq \varprojlim \text{Gal}(\mathbb{F}_{q^i}/\mathbb{F}_q).$$

Proof. For each $i \in \mathbb{N}$, we have a homomorphism $\mathrm{Gal}(\overline{\mathbb{F}_q}/\mathbb{F}_q) \to \mathrm{Gal}(\mathbb{F}_{q^i}/\mathbb{F}_q)$ obtained by restriction. These together yield a homomorphism

$$\theta : \mathrm{Gal}(\overline{\mathbb{F}_q}/\mathbb{F}_q) \to \prod_{i \in \mathbb{N}} \mathrm{Gal}(\mathbb{F}_{q^i}/\mathbb{F}_q).$$

It is clear that the image of θ is contained in $\lim_{\leftarrow} \mathrm{Gal}(\mathbb{F}_{q^i}/\mathbb{F}_q)$. We show in the following that θ is an isomorphism onto $\lim_{\leftarrow} \mathrm{Gal}(\mathbb{F}_{q^i}/\mathbb{F}_q)$.

If $\sigma \neq 1$ is in $\mathrm{Gal}(\overline{\mathbb{F}_q}/\mathbb{F}_q)$, then there exists an element $x \in \overline{\mathbb{F}_q}$ such that $\sigma(x) \neq x$. By Theorem 1.2.1, x belongs to $\mathbb{F}_{q^n}$ for some $n \in \mathbb{N}$. Now the image of σ in $\mathrm{Gal}(\mathbb{F}_{q^n}/\mathbb{F}_q)$ maps x to $\sigma(x)$, and thus $\theta(\sigma)$ is not the identity. Hence, θ is injective.

Take (σ_i) in $\lim_{\leftarrow} \mathrm{Gal}(\mathbb{F}_{q^i}/\mathbb{F}_q)$. If $x \in \overline{\mathbb{F}_q}$ and we set $\sigma(x) = \sigma_i(x)$, where $x \in \mathbb{F}_{q^i}$, then this is an unambiguous definition of a map $\sigma : \overline{\mathbb{F}_q} \to \overline{\mathbb{F}_q}$. It is easy to check that σ is an element of $\mathrm{Gal}(\overline{\mathbb{F}_q}/\mathbb{F}_q)$. Since $\theta(\sigma) = (\sigma_i)$, $\lim_{\leftarrow} \mathrm{Gal}(\mathbb{F}_{q^i}/\mathbb{F}_q)$ is the image of θ. $\qquad \square$

Corollary 1.2.6. We have

$$\mathrm{Gal}(\overline{\mathbb{F}_q}/\mathbb{F}_q) \simeq \hat{\mathbb{Z}}.$$

Proof. For each $i \in \mathbb{N}$, we can identify the group $\mathrm{Gal}(\mathbb{F}_{q^i}/\mathbb{F}_q)$ with $\mathbb{Z}/i\mathbb{Z}$ by Theorem 1.1.5(ii). Under this identification, the family of homomorphisms in Example 1.2.4 coincides with that in Example 1.2.3. Thus, the desired result follows from Theorem 1.2.5. $\qquad \square$

It is another direct consequence of Theorem 1.2.5 that the restrictions of all automorphisms in $\mathrm{Gal}(\overline{\mathbb{F}_q}/\mathbb{F}_q)$ to $\mathbb{F}_{q^m}$ give all automorphisms in $\mathrm{Gal}(\mathbb{F}_{q^m}/\mathbb{F}_q)$, that is, we obtain the following result.

Corollary 1.2.7. For every integer $m \geq 1$, we have

$$\mathrm{Gal}(\mathbb{F}_{q^m}/\mathbb{F}_q) = \{\sigma|_{\mathbb{F}_{q^m}} : \sigma \in \mathrm{Gal}(\overline{\mathbb{F}_q}/\mathbb{F}_q)\}.$$

For each $i \in \mathbb{N}$, let $\pi_i \in \mathrm{Gal}(\mathbb{F}_{q^i}/\mathbb{F}_q)$ be the automorphism $\pi_i : x \mapsto x^q$. Then the element (π_i) is in $\lim_{\leftarrow} \mathrm{Gal}(\mathbb{F}_{q^i}/\mathbb{F}_q)$. This yields an automorphism in $\mathrm{Gal}(\overline{\mathbb{F}_q}/\mathbb{F}_q)$. We call it the *Frobenius (automorphism)* of $\overline{\mathbb{F}_q}/\mathbb{F}_q$, denoted by π. It is clear that $\pi(x) = x^q$ for all $x \in \overline{\mathbb{F}_q}$ and that the restriction of π to $\mathbb{F}_{q^i}$ is π_i, the *Frobenius (automorphism)* of $\mathbb{F}_{q^i}/\mathbb{F}_q$.

1.3 Irreducible Polynomials

Let $\alpha \in \overline{\mathbb{F}_q}$ and $\sigma \in \mathrm{Gal}(\overline{\mathbb{F}_q}/\mathbb{F}_q)$. The element $\sigma(\alpha)$ is called a *conjugate* of α with respect to $\mathbb{F}_q$.

Lemma 1.3.1. The set of conjugates of an element $\alpha \in \overline{\mathbb{F}_q}$ with respect to $\mathbb{F}_q$ is equal to $\{\pi^i(\alpha) : i = 0, 1, 2, \ldots\}$, where $\pi \in \mathrm{Gal}(\overline{\mathbb{F}_q}/\mathbb{F}_q)$ is the Frobenius automorphism.

Proof. Let $\sigma \in \mathrm{Gal}(\overline{\mathbb{F}_q}/\mathbb{F}_q)$. There exists an integer $m \geq 1$ such that α is an element of $\mathbb{F}_{q^m}$. Then the restrictions $\sigma|_{\mathbb{F}_{q^m}}$ and $\pi|_{\mathbb{F}_{q^m}}$ are both elements of $\mathrm{Gal}(\mathbb{F}_{q^m}/\mathbb{F}_q)$. Moreover, $\pi|_{\mathbb{F}_{q^m}}$ is a generator of $\mathrm{Gal}(\mathbb{F}_{q^m}/\mathbb{F}_q)$. Thus, $\sigma|_{\mathbb{F}_{q^m}} = (\pi|_{\mathbb{F}_{q^m}})^i$ for some $i \geq 0$. Hence, $\sigma(\alpha) = \sigma|_{\mathbb{F}_{q^m}}(\alpha) = (\pi|_{\mathbb{F}_{q^m}})^i(\alpha) = \pi^i(\alpha)$. $\qquad\square$

Proposition 1.3.2. All distinct conjugates of an element $\alpha \in \overline{\mathbb{F}_q}$ with respect to $\mathbb{F}_q$ are $\alpha, \pi(\alpha), \ldots, \pi^{m-1}(\alpha)$, where m is the least positive integer such that $\mathbb{F}_{q^m}$ contains α, that is, m is such that $\mathbb{F}_{q^m} = \mathbb{F}_q(\alpha)$.

Proof. The restriction $\pi|_{\mathbb{F}_{q^m}}$ of π to $\mathbb{F}_{q^m}$ has order m since it is a generator of $\mathrm{Gal}(\mathbb{F}_{q^m}/\mathbb{F}_q)$. Hence, $\pi^m(\alpha) = (\pi|_{\mathbb{F}_{q^m}})^m(\alpha) = \alpha$. This implies that $\alpha, \pi(\alpha), \ldots, \pi^{m-1}(\alpha)$ yield all conjugates of α. It remains to show that they are pairwise distinct. Suppose that $\pi^n(\alpha) = \alpha$ for some $n \geq 1$. Then it is clear that $\pi^n(\beta) = \beta$ for all $\beta \in \mathbb{F}_q(\alpha)$, that is, $\beta^{q^n} - \beta = 0$ for all elements $\beta \in \mathbb{F}_{q^m}$. Thus, the polynomial $x^{q^n} - x$ has at least q^m roots. Hence, $n \geq m$. This implies that $\alpha, \pi(\alpha), \ldots, \pi^{m-1}(\alpha)$ are pairwise distinct. $\qquad\square$

Corollary 1.3.3. All distinct conjugates of an element $\alpha \in \overline{\mathbb{F}_q}$ with respect to $\mathbb{F}_q$ are $\alpha, \alpha^q, \alpha^{q^2}, \ldots, \alpha^{q^{m-1}}$, where m is the least positive integer such that $\mathbb{F}_{q^m}$ contains α, that is, m is such that $\mathbb{F}_{q^m} = \mathbb{F}_q(\alpha)$.

Proof. This follows from Proposition 1.3.2 and the fact that $\pi(\alpha) = \alpha^q$. $\qquad\square$

By field theory, all conjugates of α with respect to $\mathbb{F}_q$ form the set of all roots of the minimal polynomial of α over $\mathbb{F}_q$. Hence, we get the following result.

Corollary 1.3.4. Let f be an irreducible polynomial over $\mathbb{F}_q$ of degree m and let $\alpha \in \overline{\mathbb{F}_q}$ be a root of f. Then $\alpha, \alpha^q, \alpha^{q^2}, \ldots, \alpha^{q^{m-1}}$ are all distinct roots of f, and moreover $\mathbb{F}_{q^m} = \mathbb{F}_q(\alpha)$.

From the above result we obtain that all roots of an irreducible polynomial f over $\mathbb{F}_q$ are simple and that $\mathbb{F}_{q^m}$ is the splitting field of f over $\mathbb{F}_q$, where $m = \deg(f)$.

Lemma 1.3.5. A monic irreducible polynomial $f(x)$ of degree m over $\mathbb{F}_q$ divides $x^{q^n} - x$ if and only if m divides n.

Proof. Let $\alpha \in \overline{\mathbb{F}_q}$ be a root of $f(x)$. Then we have $\mathbb{F}_{q^m} = \mathbb{F}_q(\alpha)$ by Corollary 1.3.4. If m divides n, then $\mathbb{F}_{q^m}$ is a subfield of $\mathbb{F}_{q^n}$ by Theorem 1.1.5(iii). From Proposition 1.1.1(iii) we get $\beta^{q^n} - \beta = 0$ for all $\beta \in \mathbb{F}_{q^n}$. In particular, $\alpha^{q^n} - \alpha = 0$. Hence, the minimal polynomial $f(x)$ of α over $\mathbb{F}_q$ divides $x^{q^n} - x$.

If $f(x)$ divides $x^{q^n} - x$, then $\alpha^{q^n} - \alpha = 0$. Hence, $\alpha \in \mathbb{F}_{q^n}$ by the existence part of the proof of Theorem 1.1.2. Now $\mathbb{F}_{q^m} = \mathbb{F}_q(\alpha) \subseteq \mathbb{F}_{q^n}$ and our desired result follows from Theorem 1.1.5(iii). $\square$

Since $x^{q^n} - x$ has no multiple roots, we know from Lemma 1.3.5 that the product of all monic irreducible polynomials over $\mathbb{F}_q$ whose degrees divide n is equal to $x^{q^n} - x$. From this we obtain the number of monic irreducible polynomials over $\mathbb{F}_q$ of given degree, as stated in the following theorem.

Theorem 1.3.6. Let $I_q(n)$ be the number of monic irreducible polynomials over $\mathbb{F}_q$ of fixed degree $n \geq 1$. Then

$$I_q(n) = \frac{1}{n} \sum_{d \mid n} \mu(d) q^{n/d},$$

where the sum is over all positive integers d dividing n and μ is the Möbius function on $\mathbb{N}$ defined by

$$\mu(d) = \begin{cases} 1 & \text{if } d = 1, \\ (-1)^r & \text{if } d \text{ is the product of } r \text{ distinct primes}, \\ 0 & \text{otherwise}. \end{cases}$$

Proof. Since the product of all monic irreducible polynomials over $\mathbb{F}_q$ whose degrees divide n is equal to $x^{q^n} - x$, we obtain the identity

$$q^n = \sum_{d|n} d I_q(d)$$

by comparing degrees. Applying the Möbius inversion formula (e.g., see [72, p. 92]), we get the desired result. $\square$

1.4 Trace and Norm

In this section, we discuss two maps from the field $\mathbb{F}_{q^m}$ to the field $\mathbb{F}_q$: trace and norm.

Definition 1.4.1. The *trace* map $\mathrm{Tr}_{\mathbb{F}_{q^m}/\mathbb{F}_q}$ from $\mathbb{F}_{q^m}$ to $\mathbb{F}_q$ is defined to be

$$\sum_{\sigma \in G} \sigma,$$

where $G := \mathrm{Gal}(\mathbb{F}_{q^m}/\mathbb{F}_q)$, that is, for any $\alpha \in \mathbb{F}_{q^m}$, we put

$$\mathrm{Tr}_{\mathbb{F}_{q^m}/\mathbb{F}_q}(\alpha) = \sum_{\sigma \in G} \sigma(\alpha).$$

If there is no confusion, we simply denote the map by Tr.

For any $\tau \in \mathrm{Gal}(\mathbb{F}_{q^m}/\mathbb{F}_q)$ and $\alpha \in \mathbb{F}_{q^m}$, we have

$$\tau(\mathrm{Tr}(\alpha)) = \tau\left(\sum_{\sigma \in G} \sigma(\alpha)\right) = \sum_{\sigma \in G} (\tau\sigma)(\alpha) = \sum_{\sigma \in G} \sigma(\alpha) = \mathrm{Tr}(\alpha).$$

Thus indeed, Tr is a map from $\mathbb{F}_{q^m}$ to $\mathbb{F}_q$. Furthermore, the trace map has the following properties.

Proposition 1.4.2.

(i) $\mathrm{Tr}(\alpha + \beta) = \mathrm{Tr}(\alpha) + \mathrm{Tr}(\beta)$ for all $\alpha, \beta \in \mathbb{F}_{q^m}$.
(ii) $\mathrm{Tr}(a\alpha) = a\mathrm{Tr}(\alpha)$ for all $\alpha \in \mathbb{F}_{q^m}$ and $a \in \mathbb{F}_q$.

(iii) $\text{Tr}(\sigma(\alpha)) = \text{Tr}(\alpha)$ for all $\sigma \in \text{Gal}(\mathbb{F}_{q^m}/\mathbb{F}_q)$ and $\alpha \in \mathbb{F}_{q^m}$. In particular, $\text{Tr}(\pi(\alpha)) = \text{Tr}(\alpha^q) = \text{Tr}(\alpha)$, where $\pi \in \text{Gal}(\mathbb{F}_{q^m}/\mathbb{F}_q)$ is the Frobenius.

The proof of the above proposition easily follows from the definition of the trace map. We glean from (i) and (ii) of the above proposition that Tr is a linear transformation when we view $\mathbb{F}_{q^m}$ and $\mathbb{F}_q$ as vector spaces over $\mathbb{F}_q$.

Since π is a generator of the cyclic group $G = \text{Gal}(\mathbb{F}_{q^m}/\mathbb{F}_q)$ of order m, we have $\text{Tr} = \sum_{\sigma \in G} \sigma = \sum_{i=0}^{m-1} \pi^i$. Hence, for any $\alpha \in \mathbb{F}_{q^m}$, we have

$$\text{Tr}(\alpha) = \sum_{i=0}^{m-1} \pi^i(\alpha) = \alpha + \alpha^q + \alpha^{q^2} + \cdots + \alpha^{q^{m-1}}.$$

Theorem 1.4.3.

(i) The trace map is surjective. Thus, the kernel of Tr is a vector space of dimension $m-1$ over $\mathbb{F}_q$.
(ii) An element α of $\mathbb{F}_{q^m}$ satisfies $\text{Tr}(\alpha) = 0$ if and only if $\alpha = \pi(\beta) - \beta = \beta^q - \beta$ for some $\beta \in \mathbb{F}_{q^m}$.

Proof.

(i) An element α of $\mathbb{F}_{q^m}$ is in the kernel of Tr if and only if α is a root of the polynomial $x + x^q + x^{q^2} + \cdots + x^{q^{m-1}}$. Thus, the kernel of Tr contains at most q^{m-1} elements. Therefore, there are at least $q^m/q^{m-1} = q$ elements in the image of Tr. Since the image of Tr is a subset of $\mathbb{F}_q$, we conclude that the image is the same as $\mathbb{F}_q$. As the dimension of the $\mathbb{F}_q$-linear space $\mathbb{F}_{q^m}$ is m, the dimension of the kernel is equal to $m-1$.
(ii) Consider the $\mathbb{F}_q$-linear map $\phi : \gamma \in \mathbb{F}_{q^m} \mapsto \pi(\gamma) - \gamma$. By Proposition 1.4.2(iii), the image $\text{Im}(\phi)$ is contained in the kernel of Tr. Now $\phi(\gamma) = 0$ if and only if $\pi(\gamma) = \gamma$. This is equivalent to γ being an element of $\mathbb{F}_q$. Hence, the kernel of ϕ is $\mathbb{F}_q$. So $\text{Im}(\phi)$ contains $q^m/q = q^{m-1}$ elements. This implies that $\text{Im}(\phi)$ is the same as the kernel of Tr. Therefore, $\text{Tr}(\alpha) = 0$ if and only if $\alpha \in \text{Im}(\phi)$, that is, there exists an element $\beta \in \mathbb{F}_{q^m}$ such that $\alpha = \phi(\beta) = \pi(\beta) - \beta$. $\square$

Definition 1.4.4. The *norm* map $\mathrm{Nm}_{\mathbb{F}_{q^m}/\mathbb{F}_q}$ from $\mathbb{F}_{q^m}$ to $\mathbb{F}_q$ is defined to be

$$\prod_{\sigma \in G} \sigma,$$

where $G := \mathrm{Gal}(\mathbb{F}_{q^m}/\mathbb{F}_q)$, that is, for any $\alpha \in \mathbb{F}_{q^m}$, we put

$$\mathrm{Nm}_{\mathbb{F}_{q^m}/\mathbb{F}_q}(\alpha) = \prod_{\sigma \in G} \sigma(\alpha).$$

If there is no confusion, we simply denote the map by Nm.

For any $\tau \in \mathrm{Gal}(\mathbb{F}_{q^m}/\mathbb{F}_q)$ and $\alpha \in \mathbb{F}_{q^m}$, we have

$$\tau(\mathrm{Nm}(\alpha)) = \tau\left(\prod_{\sigma \in G} \sigma(\alpha)\right) = \prod_{\sigma \in G}(\tau\sigma)(\alpha) = \prod_{\sigma \in G} \sigma(\alpha) = \mathrm{Nm}(\alpha).$$

Thus indeed, Nm is a map from $\mathbb{F}_{q^m}$ to $\mathbb{F}_q$. Furthermore, the norm map has the following properties.

Proposition 1.4.5.

(i) $\mathrm{Nm}(\alpha \cdot \beta) = \mathrm{Nm}(\alpha) \cdot \mathrm{Nm}(\beta)$ for all $\alpha, \beta \in \mathbb{F}_{q^m}$.

(ii) $\mathrm{Nm}(a\alpha) = a^m \mathrm{Nm}(\alpha)$ for all $\alpha \in \mathbb{F}_{q^m}$ and $a \in \mathbb{F}_q$.

(iii) $\mathrm{Nm}(\sigma(\alpha)) = \mathrm{Nm}(\alpha)$ for all $\sigma \in \mathrm{Gal}(\mathbb{F}_{q^m}/\mathbb{F}_q)$ and $\alpha \in \mathbb{F}_{q^m}$. In particular, $\mathrm{Nm}(\pi(\alpha)) = \mathrm{Nm}(\alpha^q) = \mathrm{Nm}(\alpha)$, where $\pi \in \mathrm{Gal}(\mathbb{F}_{q^m}/\mathbb{F}_q)$ is the Frobenius.

The proof of the above proposition easily follows from the definition of the norm map. We obtain from (i) of the above proposition that Nm is a group homomorphism from $\mathbb{F}_{q^m}^*$ to $\mathbb{F}_q^*$.

Since π is a generator of the cyclic group $G = \mathrm{Gal}(\mathbb{F}_{q^m}/\mathbb{F}_q)$ of order m, we have $\mathrm{Nm} = \prod_{\sigma \in G} \sigma = \prod_{i=0}^{m-1} \pi^i$. Hence, for any $\alpha \in \mathbb{F}_{q^m}$, we have

$$\mathrm{Nm}(\alpha) = \prod_{i=0}^{m-1} \pi^i(\alpha) = \alpha \cdot \alpha^q \cdot \alpha^{q^2} \cdots \alpha^{q^{m-1}}.$$

Theorem 1.4.6.

(i) The norm map is an epimorphism from $\mathbb{F}_{q^m}^*$ to $\mathbb{F}_q^*$. Thus, the kernel of Nm is a cyclic group of order $(q^m - 1)/(q - 1)$.

(ii) An element α of $\mathbb{F}_{q^m}$ satisfies $\mathrm{Nm}(\alpha) = 1$ if and only if $\alpha = \pi(\beta)/\beta = \beta^{q-1}$ for some $\beta \in \mathbb{F}_{q^m}^*$.

Proof. Using similar arguments as in the proof of Theorem 1.4.3 and replacing Tr by Nm, we obtain the desired results. $\square$

1.5 Function Fields of One Variable

For a given field k, a *function field* over k is a field extension F of k such that there is at least one element $x \in F$ that is transcendental over k. The field k is called a *constant field* of F.

For a function field F over k, we consider the set k_1 of elements of F that are algebraic over k. It is clear that k_1 is a field since sums, products, and inverses of algebraic elements over k are again algebraic over k. Hence, we have the chain of fields $k \subseteq k_1 \subseteq F$.

Lemma 1.5.1. Let F be a function field over k and let k_1 be the set of elements of F that are algebraic over k. Then:

(i) F is also a function field over k_1;

(ii) k_1 is algebraically closed in F, that is, all elements in $F \setminus k_1$ are transcendental over k_1.

Proof.

(i) Let $x \in F$ be a transcendental element over k. We will show that x is also transcendental over k_1. Suppose that x were algebraic over k_1. Let $\sum_{i=0}^{n} c_i T^i$ with $c_i \in k_1$ be the minimal polynomial of x over k_1. Then x is algebraic over $k(c_0, c_1, \ldots, c_n)$. Hence, $k(c_0, c_1, \ldots, c_n, x)$ is a finite extension of $k(c_0, c_1, \ldots, c_n)$. Since all c_i are algebraic over k, it follows that $k(c_0, c_1, \ldots, c_n)$ is also a finite extension of k. Therefore, $[k(c_0, c_1, \ldots, c_n, x) : k] = [k(c_0, c_1, \ldots, c_n, x) : k(c_0, c_1, \ldots, c_n)] \cdot [k(c_0, c_1, \ldots, c_n) : k] < \infty$. Thus $[k(x) : k] \le [k(c_0, c_1, \ldots, c_n, x) : k] < \infty$. This implies that x is algebraic over k, a contradiction.

(ii) It suffices to show that if $\alpha \in F$ is algebraic over k_1, then α is an element of k_1. Let $\alpha \in F$ be an algebraic element over k_1 and let $\sum_{i=0}^{n} c_i T^i$ with $c_i \in k_1$ be its minimal polynomial over k_1. With the same arguments as in the proof of (i), we can show that $[k(\alpha) : k] < \infty$, that is, α is algebraic over k. Hence, α is in k_1.

□

For a function field F over k, we say that k is algebraically closed in F if k is the same as $k_1 = \{\alpha \in F : \alpha \text{ is algebraic over } k\}$. In this case, we call k the *full constant field* of F. From now on we always mean that F is a function field over k with full constant field k whenever we write F/k. We will now concentrate on algebraic function fields of one variable, which are defined as follows.

Definition 1.5.2. The function field F/k is an *algebraic function field of one variable* over k if there exists a transcendental element $x \in F$ over k such that F is a finite extension of the rational function field $k(x)$. If in addition the full constant field k is finite, then F/k is called a *global function field*.

The study of algebraic function fields of one variable has a long history. Classical books on this topic include those by Chevalley [18] and Deuring [25]. The more recent book by Stichtenoth [117] puts a special emphasis on global function fields.

In the rest of this section, F/k will always denote an algebraic function field of one variable. We develop the theory of algebraic function fields of one variable by starting from the concept of a valuation. We add ∞ to the field $\mathbb{R}$ of real numbers to form the set $\mathbb{R} \cup \{\infty\}$, and we put $\infty + \infty = \infty + c = c + \infty = \infty$ for any $c \in \mathbb{R}$. We agree that $c < \infty$ for any $c \in \mathbb{R}$.

Definition 1.5.3. A *valuation* of F/k is a map $\nu : F \to \mathbb{R} \cup \{\infty\}$ satisfying the following conditions:

1. $\nu(z) = \infty$ if and only if $z = 0$;
2. $\nu(yz) = \nu(y) + \nu(z)$ for all $y, z \in F$;
3. $\nu(y + z) \geq \min(\nu(y), \nu(z))$ for all $y, z \in F$;
4. $\nu(F^*) \neq \{0\}$;
5. $\nu(\alpha) = 0$ for all $\alpha \in k^*$.

Remark 1.5.4.

(i) Condition (3) is called the *triangle inequality*. In fact, we have a stronger result called the *strict triangle inquality*, which says that

$$v(y + z) = \min{(v(y), v(z))} \tag{1.1}$$

whenever $v(y) \neq v(z)$. In order to show (1.1), we can assume that $v(y) < v(z)$. Suppose that $v(y + z) \neq \min{(v(y), v(z))}$. Then $v(y + z) > \min{(v(y), v(z))} = v(y)$ by condition (3). Thus $v(y) = v((y+z) - z) \geq \min{(v(y+z), v((-1)\cdot z))} = \min{(v(y+z), v(z))} > v(y)$, a contradiction.

(ii) If k is finite, then condition (5) follows from the other conditions in Definition 1.5.3. Note that if $k = \mathbb{F}_q$, then $\alpha^{q-1} = 1$ for all $\alpha \in k^*$ by Proposition 1.1.1(iii), and so $0 = v(1) = v(\alpha^{q-1}) = (q-1)v(\alpha)$, which yields $v(\alpha) = 0$.

If the image $v(F^*)$ is a discrete set in $\mathbb{R}$, then v is called *discrete*. If $v(F^*) = \mathbb{Z}$, then v is called *normalized*.

Example 1.5.5. Consider the rational function field $k(x)$. The full constant field of $k(x)$ is k since it is easily seen that a nonconstant rational function in $k(x)$ cannot be algebraic over k. Let $p(x)$ be a monic irreducible polynomial in $k[x]$. Let the map $v_{p(x)} : k(x) \rightarrow \mathbb{Z} \cup \{\infty\}$ be defined as follows:

(i) for a nonzero polynomial $f(x) \in k[x]$ with $p^m(x) \| f(x)$, that is, $p^m(x) | f(x)$ and $p^{m+1}(x) \nmid f(x)$, put $v_{p(x)}(f(x)) = m$;

(ii) for a nonzero rational function $f(x)/g(x) \in k(x)$, put $v_{p(x)}(f(x)/g(x)) = v_{p(x)}(f(x)) - v_{p(x)}(g(x))$;

(iii) put $v_{p(x)}(0) = \infty$.

It is easy to verify that $v_{p(x)}$ is a well-defined map and satisfies the conditions in Definition 1.5.3. Hence, $v_{p(x)}$ is a (discrete) normalized valuation of $k(x)$.

Besides the valuations $v_{p(x)}$ defined above, we have another (discrete) normalized valuation v_∞ of $k(x)$ defined by

$$v_\infty\left(\frac{f(x)}{g(x)}\right) = \deg(g(x)) - \deg(f(x))$$

for any two nonzero polynomials $f(x), g(x) \in k[x]$ and $v_\infty(0) = \infty$. It is easy to show that v_∞ is a well-defined valuation map.

We will see later in this section (see Theorem 1.5.8) that every normalized valuation of $k(x)$ is of the form $v_{p(x)}$ for some monic irreducible polynomial $p(x) \in k[x]$ or v_∞.

Two discrete valuations v and λ of F/k are called *equivalent* if there exists a constant $c > 0$ such that

$$v(z) = c\lambda(z) \quad \text{for all } z \in F^*.$$

Obviously, this yields an equivalence relation between discrete valuations of F/k. An equivalence class of discrete valuations of F/k is called a *place* of F/k.

If v is a discrete valuation of F/k, then $v(F^*)$ is a nonzero discrete subgroup of $(\mathbb{R}, +)$, and so we have $v(F^*) = b\mathbb{Z}$ for some positive $b \in \mathbb{R}$. Thus, there exists a uniquely determined normalized valuation of F that is equivalent to v. In other words, every place P of F/k contains a uniquely determined normalized valuation of F/k, which is denoted by v_P. Thus, we can identify places of F/k and (discrete) normalized valuations of F/k.

For a place P of F/k and an element $z \in F^*$, we say that P is a *zero* of z if $v_P(z) > 0$ and that P is a *pole* of z if $v_P(z) < 0$.

For the normalized valuation v_P of F/k we have $v_P(F^*) = \mathbb{Z}$. Thus, there exists an element $t \in F$ satisfying $v_P(t) = 1$. Such an element t is called a *local parameter* (or *uniformizing parameter*) of F at the place P.

For a place P of F/k, we set

$$\mathcal{O}_P = \{z \in F : v_P(z) \geq 0\}.$$

Using the properties of valuations, it is easy to show that $\mathcal{O}_P$ forms a subring of F with $k \subseteq \mathcal{O}_P$. We call $\mathcal{O}_P$ the *valuation ring* of the place P.

Proposition 1.5.6. The valuation ring $\mathcal{O}_P$ has a unique maximal ideal given by

$$M_P := \{z \in F : v_P(z) \geq 1\}.$$

Proof. It is trivial that M_P is an ideal of $\mathcal{O}_P$. Since $1 \in \mathcal{O}_P \setminus M_P$, we obtain that M_P is a proper ideal. It remains to show that any proper ideal J of $\mathcal{O}_P$ is contained in M_P. Take $z \in J$ and suppose that $v_P(z) = 0$. Then $v_P(z^{-1}) = -v_P(z) = 0$, and so $z^{-1} \in \mathcal{O}_P$. Thus, $1 = z^{-1}z \in J$ and, hence, $J = \mathcal{O}_P$, a contradiction. Therefore, $v_P(z) \geq 1$ and $J \subseteq M_P$.

$\square$

The ideal M_P is called the *maximal ideal* of the place P. It is, in fact, a principal ideal since $M_P = t\mathcal{O}_P$ for any local parameter t of F at P.

Remark 1.5.7. Valuation rings in F/k can be characterized purely algebraically as the maximal proper Noetherian subrings of F/k, or also as the maximal principal ideal domains properly contained in F/k (see [105, Section 2.1]). A place of F/k can then be defined as the maximal ideal of some valuation ring in F/k. This alternative approach to the concept of a place is presented in the book of Stichtenoth [117].

We now determine all discrete valuations of the rational function field $k(x)$. The valuations $v_{p(x)}$ and v_∞ are defined as in Example 1.5.5.

Theorem 1.5.8. Every discrete valuation of the rational function field $k(x)$ is equivalent to either $v_{p(x)}$ for some monic irreducible polynomial $p(x)$ in $k[x]$ or to v_∞.

Proof. Let v be a discrete valuation of $k(x)$ and let P be its corresponding place. We distinguish two cases according to the value of $v(x)$.

Case 1: $v(x) \geq 0$. Then $k[x] \subseteq \mathcal{O}_P$ and $v(k(x)^*) \neq \{0\}$ by the properties of a valuation, and so $J := k[x] \cap M_P$ is a nonzero ideal of $k[x]$. Furthermore, $J \neq k[x]$ since $1 \notin J$. Since M_P is a prime ideal of $\mathcal{O}_P$, it follows that J is a prime ideal of $k[x]$. Consequently, there exists a monic irreducible $p(x) \in k[x]$ such that J is the principal ideal $(p(x))$. In particular, $c := v(p(x)) > 0$. If $h(x) \in k[x]$ is not divisible by $p(x)$, then $h(x) \notin M_P$, and so $v(h(x)) = 0$. Thus, if we write a nonzero $r(x) \in k(x)$ in the form

$$r(x) = p(x)^m \frac{f(x)}{g(x)}$$

with $m \in \mathbb{Z}$ and $f(x), g(x) \in k[x]$ not divisible by $p(x)$, then

$$v(r(x)) = mv(p(x)) = cv_{p(x)}(r(x)),$$

and so v is equivalent to $v_{p(x)}$.

Case 2: $v(x) < 0$. Then $c := v(x^{-1}) > 0$ and $x^{-1} \in \mathsf{M}_P$. Take any nonzero $f(x) \in k[x]$ of degree d, say. Then

$$f(x) = \sum_{i=0}^{d} \alpha_i x^i = x^d \sum_{i=0}^{d} \alpha_i x^{i-d} = x^d \sum_{i=0}^{d} \alpha_{d-i} x^{-i}$$

with all $\alpha_i \in k$. Furthermore,

$$\sum_{i=0}^{d} \alpha_{d-i} x^{-i} = \alpha_d + \sum_{i=1}^{d} \alpha_{d-i} x^{-i} = \alpha_d + s(x)$$

with $s(x) \in \mathsf{M}_P$. Since $\alpha_d \neq 0$, we have $v(\alpha_d) = 0$, and so

$$v\left(\sum_{i=0}^{d} \alpha_{d-i} x^{-i}\right) = 0$$

by the strict triangle inequality (see Remark 1.5.4(i)). It follows that

$$v(f(x)) = v(x^d) = -dv(x^{-1}) = cv_\infty(f(x)),$$

and so v is equivalent to v_∞.   □

Remark 1.5.9. If we write $\mathcal{O}_v = \{z \in F : v(z) \geq 0\}$ and $\mathsf{M}_v = \{z \in F : v(z) > 0\}$ for an arbitrary valuation v of F/k and let $\mathcal{O}_v$ and M_v play the roles of $\mathcal{O}_P$ and M_P, respectively, in the proof of Theorem 1.5.8, then the argument in the proof goes through. Thus, this proof demonstrates that every valuation of $k(x)$ is automatically discrete.

Theorem 1.5.8 shows that there are exactly two types of places of the rational function field $k(x)$: (i) the *finite places* containing some valuation $v_{p(x)}$; (ii) the *infinite place* containing the valuation v_∞. Note that the valuations $v_{p(x)}$ and v_∞ are nonequivalent since $v_{p(x)}(p(x)) = 1$, whereas $v_\infty(p(x)) < 0$. Furthermore, $v_{p(x)}$ and $v_{q(x)}$ are nonequivalent for distinct monic irreducible $p(x), q(x) \in k[x]$ since $v_{p(x)}(p(x)) = 1$, whereas $v_{q(x)}(p(x)) = 0$. Thus, there is a one-to-one correspondence between the finite places of $k(x)$ and the monic irreducible polynomials in $k[x]$. We may use this correspondence to speak, for instance, of the place $p(x)$ of $k(x)$ when we mean the place containing the valuation $v_{p(x)}$. To summarize, the set of distinct places of $k(x)$ can be identified with the set

$$\{p(x) \in k[x] : p(x) \text{ monic irreducible}\} \cup \{\infty\}.$$

Definition 1.5.10. Let P be a place of the algebraic function field F/k of one variable over k. The field $F_P := \mathcal{O}_P/M_P$ is called the *residue class field* of P. The canonical ring homomorphism

$$z \in \mathcal{O}_P \mapsto z(P) := z + M_P \in F_P$$

is called the *residue class map* of P.

Example 1.5.11. Consider again the rational function field $k(x)$. If $p(x) \in k[x]$ is monic irreducible, then for the finite place $p(x)$ of $k(x)$, we have

$$\mathcal{O}_{p(x)} = \left\{ \frac{f(x)}{g(x)} : f(x), g(x) \in k[x], \ p(x) \nmid g(x) \right\},$$

$$M_{p(x)} = \left\{ \frac{f(x)}{g(x)} : f(x), g(x) \in k[x], \ p(x)|f(x), \ p(x) \nmid g(x) \right\}.$$

For any $h(x) \in k[x]$, we write $\overline{h(x)}$ for the residue class of $h(x)$ modulo the ideal $(p(x))$ of $k[x]$. If $p(x) \nmid h(x)$, then $\overline{h(x)} \in k[x]/(p(x))$ has a multiplicative inverse $\overline{h(x)}^{-1} \in k[x]/(p(x))$. The map $\psi_{p(x)} : \mathcal{O}_{p(x)} \to k[x]/(p(x))$ given by

$$\psi_{p(x)}\left(\frac{f(x)}{g(x)}\right) = \overline{f(x)}\ \overline{g(x)}^{-1} \quad \text{for all } \frac{f(x)}{g(x)} \in \mathcal{O}_{p(x)}$$

is well defined. Clearly, $\psi_{p(x)}$ is a surjective ring homomorphism with kernel $M_{p(x)}$, and so the residue class field of the place $p(x)$ is isomorphic to $k[x]/(p(x))$. For the infinite place ∞ of $k(x)$, we have

$$\mathcal{O}_\infty = \left\{ \frac{f(x)}{g(x)} : f(x), g(x) \in k[x], \ g(x) \neq 0, \ \deg(f(x)) \leq \deg(g(x)) \right\},$$

$$M_\infty = \left\{ \frac{f(x)}{g(x)} : f(x), g(x) \in k[x], \ \deg(f(x)) < \deg(g(x)) \right\},$$

where we put as usual $\deg(0) = -\infty$. Every $r(x) \in \mathcal{O}_\infty$ can be written in the form

$$r(x) = \frac{\alpha_d x^d + \alpha_{d-1} x^{d-1} + \cdots + \alpha_0}{x^d + \beta_{d-1} x^{d-1} + \cdots + \beta_0}$$

with all $\alpha_i, \beta_j \in k$ and $d \geq 0$. The map $\psi_\infty : \mathcal{O}_\infty \to k$ given by

$$\psi_\infty(r(x)) = \alpha_d \quad \text{for all } r(x) \in \mathcal{O}_\infty$$

is well defined. It is easily seen that ψ_∞ is a surjective ring homomorphism with kernel M_∞, and so the residue class field of the place ∞ is isomorphic to k.

Theorem 1.5.12. Every valuation of an algebraic function field of one variable is discrete.

Proof. Let F/k be an algebraic function field of one variable. By Definition 1.5.2, there exists a transcendental element $x \in F$ over k such that F is a finite extension of $K := k(x)$. Let v be an arbitrary valuation of F and let ξ be the restriction of v to K. It suffices to prove that the index $[v(F^*) : \xi(K^*)]$ is finite. Since $v(F^*)$ is an infinite subgroup of $(\mathbb{R}, +)$, this shows then that $\xi(K^*) = \{0\}$ is not possible. Hence, ξ is a valuation of K, thus discrete by Remark 1.5.9, and so v is discrete.

Let $z_1, \ldots, z_n \in F^*$ be such that $v(z_1), \ldots, v(z_n)$ are in distinct cosets modulo $\xi(K^*)$. We claim that $z_1, \ldots, z_n$ are linearly independent over K. This will then show that

$$[v(F^*) : \xi(K^*)] \leq [F : K] < \infty.$$

So, suppose we had

$$\sum_{i=1}^{n} b_i z_i = 0,$$

where without loss of generality all $b_i \in K^*$. If we had $v(b_i z_i) = v(b_j z_j)$ for some i and j with $1 \leq i < j \leq n$, then

$$v(z_i) - v(z_j) = v(b_j) - v(b_i) = \xi(b_j b_i^{-1}) \in \xi(K^*),$$

a contradiction to the choice of $z_1, \ldots, z_n$. Thus, $v(b_1 z_1), \ldots, v(b_n z_n)$ are all distinct, and so the strict triangle inequality yields

$$v\left(\sum_{i=1}^{n} b_i z_i\right) = \min_{1 \leq i \leq n} \; v(b_i z_i) < \infty,$$

which is again a contradiction. $\square$

In the above proof, we have shown, in particular, that the restriction of a valuation of F/k to $k(x)$ yields a valuation of $k(x)$. Obviously, for equivalent valuations of F the restrictions are again equivalent. Thus, a place Q of F corresponds by restriction to a unique place P of $k(x)$. We say that Q *lies over* P or that P *lies under* Q. Therefore, every place of F lies either over a place of $k(x)$ corresponding to a monic irreducible polynomial in $k[x]$ or over the infinite place of $k(x)$.

Theorem 1.5.13. *The residue class field of every place of F/k is a finite extension (of an isomorphic copy) of k.*

Proof. Let Q be a place of F that lies over the place P of $K := k(x)$ with x as in Definition 1.5.2. Let $R_Q := \mathcal{O}_Q/\mathsf{M}_Q$ and $R_P := \mathcal{O}_P/\mathsf{M}_P$ be the corresponding residue class fields and note that $\mathcal{O}_P \subseteq \mathcal{O}_Q$. The map $\rho : R_P \to R_Q$ given by

$$\rho(b + \mathsf{M}_P) = b + \mathsf{M}_Q \quad \text{for all } b \in \mathcal{O}_P$$

is well defined since $\mathsf{M}_P \subseteq \mathsf{M}_Q$. It is clear that ρ is an injective ring homomorphism, and so R_Q contains the isomorphic copy $\rho(R_P)$ of R_P as a subfield.

Let $z_1, \ldots, z_n \in \mathcal{O}_Q$ be such that $z_1 + \mathsf{M}_Q, \ldots, z_n + \mathsf{M}_Q$ are linearly independent over $\rho(R_P)$. We claim that $z_1, \ldots, z_n$ are linearly independent over K. This will then show that

$$[R_Q : \rho(R_P)] \leq [F : K] < \infty.$$

Since R_P is a finite extension (of an isomorphic copy) of k (see Example 1.5.11), this proves the theorem. So, suppose we had

$$\sum_{i=1}^{n} b_i z_i = 0$$

with $b_1, \ldots, b_n \in K$ not all 0. Without loss of generality

$$\nu_P(b_1) = \min_{1 \le i \le n} \nu_P(b_i).$$

Then $b_1 \ne 0$ and

$$z_1 + \sum_{i=2}^{n} b_i b_1^{-1} z_i = 0.$$

By the condition on $\nu_P(b_1)$, we have $b_i b_1^{-1} \in \mathcal{O}_P$ for $2 \le i \le n$. Passing to the residue classes modulo M_Q, we get

$$(z_1 + M_Q) + \sum_{i=2}^{n} \rho(b_i b_1^{-1} + M_P)(z_i + M_Q) = 0 + M_Q,$$

a contradiction to the choice of $z_1, \ldots, z_n$.   □

In view of Theorem 1.5.13, the following definition is meaningful.

Definition 1.5.14. The *degree* $\deg(P)$ of a place P of F/k is defined to be the degree of the residue class field of P over k. A place of F/k of degree 1 is also called a *rational place* of F/k.

Example 1.5.15. Let $F = k(x)$ be the rational function field over k. As we noted in Example 1.5.5, the full constant field of F is k. By Example 1.5.11, the degree of a finite place $p(x)$ of F is equal to the degree of the polynomial $p(x)$ and the degree of the infinite place of F is equal to 1. If $k = \mathbb{F}_q$, then the rational function field F has thus exactly $q + 1$ rational places.

Next we prove the approximation theorem for valuations of algebraic function fields of one variable. The following two preparatory results are needed for the proof.

Lemma 1.5.16. If P_1 and P_2 are two distinct places of F/k, then there exists an element $z \in F$ such that $\nu_{P_1}(z) > 0$ and $\nu_{P_2}(z) \le 0$.

Proof. First consider any $u \in F$ with $v_{P_2}(u) = 0$. If $v_{P_1}(u) \neq 0$, then either $z = u$ or $z = u^{-1}$ works. Thus, we are left with the case where $v_{P_2}(u) = 0$ always implies $v_{P_1}(u) = 0$.

Let $y \in F^*$ be arbitrary and let $t \in F$ be a local parameter at P_2. Then we can write $y = t^n u$ with $n \in \mathbb{Z}$ and $v_{P_2}(u) = 0$. It follows that $v_{P_1}(u) = 0$, and so $v_{P_1}(y) = n v_{P_1}(t)$. Since v_{P_1} is normalized, we must have $v_{P_1}(t) = \pm 1$. If $v_{P_1}(t) = 1$, then $v_{P_1}(y) = n = v_{P_2}(y)$ for all $y \in F^*$, a contradiction to $P_1 \neq P_2$. Thus, $v_{P_1}(t) = -1$, and then we take $z = t^{-1}$.   □

Lemma 1.5.17. If $P_1, \ldots, P_n$ are distinct places of F/k, then there exists an element $z \in F$ such that $v_{P_1}(z) > 0$ and $v_{P_i}(z) < 0$ for $2 \leq i \leq n$.

Proof. We proceed by induction on n. For $n = 2$, we apply Lemma 1.5.16 and we get $w \in F$ with $v_{P_1}(w) > 0$ and $v_{P_2}(w) \leq 0$ as well as $y \in F$ with $v_{P_2}(y) > 0$ and $v_{P_1}(y) \leq 0$. Then we take $z = wy^{-1}$.

Assume that the lemma is true for $n - 1$ distinct places for some $n \geq 3$. By this hypothesis, there exists $w \in F$ such that $v_{P_1}(w) > 0$ and $v_{P_i}(w) < 0$ for $2 \leq i \leq n - 1$. There also exists $y \in F$ with $v_{P_1}(y) > 0$ and $v_{P_n}(y) < 0$. If $v_{P_n}(w) < 0$, we take $z = w$. If $v_{P_n}(w) \geq 0$, we put $z = w + y^r$ with an integer $r \geq 1$. Then $v_{P_1}(z) > 0$, and for $2 \leq i \leq n$ we obtain

$$v_{P_i}(z) = \min\left(v_{P_i}(w), r v_{P_i}(y)\right) < 0$$

by choosing r in such a way that the strict triangle inequality applies.   □

Theorem 1.5.18 (Approximation Theorem). Let $P_1, \ldots, P_n$ be distinct places of F/k. Then for any given elements $w_1, \ldots, w_n \in F$ and integers $m_1, \ldots, m_n$, there exists an element $z \in F$ such that

$$v_{P_i}(z - w_i) = m_i \quad \text{for } 1 \leq i \leq n.$$

Proof. We first treat the special case $w_1 = 1, w_2 = \cdots = w_n = 0$, and we show the weaker result that there exists $y \in F$ with

$$v_{P_1}(y - 1) > m_1, \qquad v_{P_i}(y) > m_i \quad \text{for } 2 \leq i \leq n. \tag{1.2}$$

By Lemma 1.5.17 we get $w \in F$ with $v_{P_1}(w) > 0$ and $v_{P_i}(w) < 0$ for $2 \leq i \leq n$. Now put $y = (1 + w^s)^{-1}$ with an integer $s \geq 1$. Then for sufficiently large s we have $v_{P_1}(y-1) = v_{P_1}(-w^s(1+w^s)^{-1}) = s v_{P_1}(w) > m_1$ and $v_{P_i}(y) = -v_{P_i}(1 + w^s) = -s v_{P_i}(w) > m_i$ for $2 \leq i \leq n$.

Now let $w_1, \ldots, w_n \in F$ be arbitrary. Choose $b \in \mathbb{Z}$ such that $v_{P_i}(w_j) \geq b$ for all $1 \leq i, j \leq n$ and put $d_i = m_i - b$ for $1 \leq i \leq n$. By (1.2) we obtain $y_1, \ldots, y_n \in F$ such that for $1 \leq i, j \leq n$ we have

$$v_{P_i}(y_i - 1) > d_i, \qquad v_{P_i}(y_j) > d_i \quad \text{for } j \neq i.$$

Let $y = \sum_{j=1}^{n} w_j y_j$. Then we can write

$$y - w_i = w_i(y_i - 1) + \sum_{\substack{j=1 \\ j \neq i}}^{n} w_j y_j,$$

and so

$$v_{P_i}(y - w_i) > b + d_i = m_i \quad \text{for } 1 \leq i \leq n \qquad (1.3)$$

by the triangle inequality.

Finally, for each $i = 1, \ldots, n$, choose $u_i \in F$ such that $v_{P_i}(u_i) = m_i$. By (1.3) there exists $z \in F$ with

$$v_{P_i}(z - (u_i + w_i)) > m_i \quad \text{for } 1 \leq i \leq n.$$

Then

$$v_{P_i}(z - w_i) = \min(v_{P_i}(z - u_i - w_i), v_{P_i}(u_i)) = m_i$$

for $1 \leq i \leq n$ by the strict triangle inequality. $\qquad \square$

Let P be a place of F/k and let $t \in F$ be a local parameter of F at P. Given an element $z \in F$, we choose an integer r such that $v_P(z) \geq r$. Then $zt^{-r} \in \mathcal{O}_P$, and so we can put

$$a_r = (zt^{-r})(P).$$

Note that a_r is an element of the residue class field F_P of P. Next we observe that

$$v_P(z - a_r t^r) \geq r + 1,$$

and so

$$v_P(zt^{-r-1} - a_r t^{-1}) \geq 0.$$

Thus, we can put

$$a_{r+1} = (zt^{-r-1} - a_r t^{-1})(P).$$

Then $a_{r+1} \in F_P$ and

$$v_P(z - a_r t^r - a_{r+1} t^{r+1}) \geq r + 2.$$

Now we construct further elements $a_j \in F_P$ inductively. Assume that for an integer $m > r$ we have already obtained elements $a_r, a_{r+1}, \ldots, a_m \in F_P$ such that

$$v_P\left(z - \sum_{j=r}^{m} a_j t^j\right) \geq m + 1.$$

Then we put

$$a_{m+1} = \left(zt^{-m-1} - \sum_{j=r}^{m} a_j t^{j-m-1}\right)(P) \in F_P,$$

and this yields

$$v_P\left(z - \sum_{j=r}^{m+1} a_j t^j\right) \geq m + 2.$$

We summarize this construction in the formal expansion

$$z = \sum_{j=r}^{\infty} a_j t^j \qquad (1.4)$$

which is called the *local expansion* of z at P. Note that if $a_r \neq 0$ in (1.4), then $v_P(z) = r$.

1.6 Extensions of Valuations

Let F/k and E/k' be algebraic function fields of one variable such that $F \subseteq E$, $[E : F] < \infty$, and their full constant fields k and k' satisfy $k \subseteq k'$. Note that then $[k' : k] < \infty$. In this section we discuss the relationships between valuations, or equivalently places, of F and E. More general situations in which a valuation is extended from a given field to an extension field are treated in the books of Ribenboim [105] and Weiss [129].

Lemma 1.6.1. For any valuation v of E/k', we have

$$[v(E^*) : v(F^*)] \leq [E : F] < \infty.$$

Proof. This is shown in the same way as in the special case considered in the proof of Theorem 1.5.12. $\qquad \Box$

If v is a valuation of E/k', then it is trivial that its restriction to F/k satisfies the conditions (1), (2), (3), and (5) in Definition 1.5.3. Moreover, Lemma 1.6.1 implies that the restriction also has the property (4). Thus, the restriction of v to F/k yields a valuation of F/k. Clearly, for equivalent valuations of E the restrictions are again equivalent. Hence, a place Q of E corresponds by restriction to a unique place P of F. We extend a terminology introduced in Section 1.5 and say that Q *lies over* P or that P *lies under* Q. Recall that v_Q denotes the normalized valuation belonging to the place Q.

Definition 1.6.2. If the place Q of E lies over the place P of F, then the *ramification index* $e(Q|P)$ of Q over P is the positive integer

$$e(Q|P) := [v_Q(E^*) : v_Q(F^*)] = [\mathbb{Z} : v_Q(F^*)].$$

We say that Q is *unramified* in the extension E/F if $e(Q|P) = 1$, that Q is *ramified* in the extension E/F if $e(Q|P) > 1$, and that Q is *totally ramified* in the extension E/F if $e(Q|P) = [E : F]$.

With v_P denoting as usual the normalized valuation belonging to the place P, Definition 1.6.2 shows that

$$v_Q(z) = e(Q|P)v_P(z) \quad \text{for all } z \in F^*.$$

If the places Q and P are as above, then their residue class fields E_Q and F_P satisfy $[E_Q : k'] < \infty$ and $[F_P : k] < \infty$ according to Theorem 1.5.13. Furthermore, the well-defined map $z(P) \mapsto z(Q)$ for $z \in \mathcal{O}_P$ provides an embedding of F_P into E_Q, and so E_Q can be viewed as a finite extension of F_P.

Definition 1.6.3. If the place Q of E lies over the place P of F, then the *relative degree* $f(Q|P)$ of Q over P is the positive integer $f(Q|P) := [E_Q : F_P]$.

According to a general principle concerning extensions of valuations (see for example, [129, Theorem 1.6.5]), if the place P of F is given and $Q_1, \ldots, Q_r$ are distinct places of E lying over P, then

$$\sum_{i=1}^{r} e(Q_i|P)f(Q_i|P) \leq [E : F]. \tag{1.5}$$

This shows, in particular, that there can be at most $[E : F]$ places of E lying over P. Note also that if one of the places of E lying over P is totally ramified in E/F, then there can be no other places of E lying over P.

Definition 1.6.4. A place P of F *splits completely* in the extension E/F if there are exactly $[E : F]$ places of E lying over P.

It follows from (1.5) that if P splits completely in E/F, then $e(Q|P) = f(Q|P) = 1$ for each place Q of E lying over P.

Given a place P of F, does there exist a place of E lying over it? There is an elegant argument in Ribenboim [105, Chapter 4, Theorem 1], which uses Zorn's lemma and shows that any valuation of a field can be extended to any algebraic extension of the field. Explicit constructions of extended valuations can be found in Ribenboim [105, Chapter 4] and Weiss [129, Chapter 2].

It is a well-known fact (see [117, Theorem III.1.11]) that if the full constant field k of F is perfect, that is, if every algebraic extension of k is separable, and if $Q_1, \ldots, Q_r$ are *all* distinct places of E lying over P, then (1.5) can be strengthened to

$$\sum_{i=1}^{r} e(Q_i|P)f(Q_i|P) = [E : F].$$ (1.6)

This provides an important relationship between ramification indices, relative degrees, and the degree of the extension E/F. Since every finite field is perfect (see, e.g., Corollary 1.3.4), (1.6) holds in particular if F (and therefore E) is a global function field.

1.7 Constant Field Extensions

Throughout this section, we are given:

(i) a global function field $F/\mathbb{F}_q$ with full constant field $\mathbb{F}_q$;
(ii) the composite field $F_n := F \cdot \mathbb{F}_{q^n}$, called a *constant field extension* of F and contained in a fixed algebraic closure of F, where n is a positive integer.

If $\mathbb{F}_{q^n} = \mathbb{F}_q(\beta)$, then we have $F_n = F(\beta)$. We recall from Theorem 1.1.5 (ii) that the Galois group $\mathrm{Gal}(\mathbb{F}_{q^n}/\mathbb{F}_q)$ is a cyclic group.

Lemma 1.7.1. With the notation above, we have:

(i) F_n/F is a cyclic extension with $[F_n : F] = n$ and $\mathrm{Gal}(F_n/F) \simeq \mathrm{Gal}(\mathbb{F}_{q^n}/\mathbb{F}_q)$;
(ii) the full constant field of F_n is $\mathbb{F}_{q^n}$.

Proof.

(i) Let $\mathbb{F}_{q^n} = \mathbb{F}_q(\beta)$ and let f be the minimal polynomial of β over $\mathbb{F}_q$. We will prove that f is irreducible over F. Suppose that f had a factorization $f = gh$ over F with g and h monic and $\deg(g) \geq 1$, $\deg(h) \geq 1$. It is obvious that all roots of g and h are elements of $\mathbb{F}_{q^n}$. Hence, from the fact that the coefficients of a monic polynomial are polynomial expressions of its roots, it follows that g and h are polynomials over $\mathbb{F}_{q^n}$. Thus, all coefficients of g and h are algebraic over $\mathbb{F}_q$. Hence, each coefficient of g and h is an element of $\mathbb{F}_q$ since $\mathbb{F}_q$ is algebraically closed in F, and we obtain a contradiction to f being irreducible over $\mathbb{F}_q$. This shows that $[F_n : F] = n$ and $\mathrm{Gal}(F_n/F) \simeq \mathrm{Gal}(\mathbb{F}_{q^n}/\mathbb{F}_q)$.

(ii) It is trivial that the full constant field of F_n contains $\mathbb{F}_{q^n}$ since $\mathbb{F}_{q^n}/\mathbb{F}_q$ is a finite extension. Let $z \in F_n$ be an algebraic element over $\mathbb{F}_{q^n}$. Then $\mathbb{F}_{q^n}(z)/\mathbb{F}_q$ is a finite extension, and so from (i) we get

$$n = [F_n : F] = [F \cdot \mathbb{F}_{q^n}(z) : F] = [\mathbb{F}_{q^n}(z) : \mathbb{F}_q].$$

Hence, $\mathbb{F}_{q^n}(z) = \mathbb{F}_{q^n}$, that is, $z \in \mathbb{F}_{q^n}$. This means that $\mathbb{F}_{q^n}$ is the full constant field of F_n.   □

By Lemma 1.7.1 we can identify $\mathrm{Gal}(F_n/F)$ with $\mathrm{Gal}(\mathbb{F}_{q^n}/\mathbb{F}_q)$ in the following way. Let $\{\beta_1 = 1, \beta_2, \ldots, \beta_n\}$ be a basis of $\mathbb{F}_{q^n}$ over $\mathbb{F}_q$. Then for any $\tau \in \mathrm{Gal}(\mathbb{F}_{q^n}/\mathbb{F}_q)$ and $x = \sum_{i=1}^n \beta_i x_i \in F_n$ with all $x_i \in F$, put

$$\tau(x) = \sum_{i=1}^n \tau(\beta_i) x_i.$$

Then τ is a Galois automorphism of F_n/F and all elements of $\mathrm{Gal}(F_n/F)$ are obtained in this way.

Theorem 1.7.2. Let $F_n = F \cdot \mathbb{F}_{q^n}$ be a constant field extension of F. Then for any place P of F and any place Q of F_n lying over P, the following holds:

(i) $e(Q|P) = 1$;
(ii) $\deg(Q) = d/\gcd(d, n)$, where $d = \deg(P)$ is the degree of P;
(iii) $f(Q|P) = n/\gcd(d, n)$;
(iv) there are exactly $\gcd(d, n)$ places of F_n lying over P.

Proof.

(i) Let $\{\beta_1, \ldots, \beta_n\}$ be a basis of $\mathbb{F}_{q^n}$ over $\mathbb{F}_q$. Then its discriminant $\Delta(\beta_1, \ldots, \beta_n)$ is an element of $\mathbb{F}_q^*$, and so

$$v_P(\Delta(\beta_1, \ldots, \beta_n)) = 0.$$

Hence, $\{\beta_1, \ldots, \beta_n\}$ is a P-integral basis of F_n/F, that is, every element in the integral closure of $\mathcal{O}_P$ in F_n can be written as a linear combination of $\beta_1, \ldots, \beta_n$ with coefficients from $\mathcal{O}_P$. The desired conclusion follows now from Dedekind's discriminant theorem (see [129, Theorem 4.8.14]).

(ii) We first prove that the residue class field R_Q of Q is the composite field $F_P \cdot \mathbb{F}_{q^n}$, where F_P is the residue class field of P. It is obvious that $F_P \cdot \mathbb{F}_{q^n}$ can be viewed as a subfield of R_Q (compare with Section 1.6). Since a basis $\{\beta_1, \ldots, \beta_n\}$ of $\mathbb{F}_{q^n}$ over $\mathbb{F}_q$ is a P-integral basis of F_n/F, we can write any element of R_Q in the form $z(Q)$, where

$$z = \sum_{i=1}^{n} z_i \beta_i$$

with $z_1, \ldots, z_n \in \mathcal{O}_P$. Hence,

$$z(Q) = \sum_{i=1}^{n} z_i(Q)\beta_i = \sum_{i=1}^{n} z_i(P)\beta_i \in F_P \cdot \mathbb{F}_{q^n},$$

showing that $R_Q = F_P \cdot \mathbb{F}_{q^n}$. This yields

$$\deg(Q) = [R_Q : \mathbb{F}_{q^n}] = [F_P \cdot \mathbb{F}_{q^n} : \mathbb{F}_{q^n}] = [\mathbb{F}_{q^d} \cdot \mathbb{F}_{q^n} : \mathbb{F}_{q^n}]$$
$$= \frac{d}{\gcd(d, n)}.$$

(iii) We have

$$f(Q|P) = [R_Q : F_P] = [F_P \cdot \mathbb{F}_{q^n} : F_P] = [\mathbb{F}_{q^d} \cdot \mathbb{F}_{q^n} : \mathbb{F}_{q^d}]$$
$$= \frac{n}{\gcd(d, n)}.$$

(iv) This follows from (i), (iii), and (1.6) □

2   Algebraic Varieties

This chapter is devoted to the fundamentals of algebraic geometry. After some preliminaries in Sections 2.1 and 2.2, we introduce in Section 2.3 the basic concept of an algebraic variety. According to a rough definition, an algebraic variety (or a variety in short) consists of the solutions of a system of (in general multivariate) polynomial equations over an algebraically closed field, and in addition a variety must satisfy a certain indecomposability condition. There are affine and projective varieties, depending on whether we work in an affine or a projective space. The discussion of function fields of varieties in Section 2.4 establishes links with the material on function fields in Chapter 1. Section 2.5 on morphisms and rational maps culminates in results which indicate that varieties can be considered equivalent if and only if their function fields are isomorphic.

A classical treatise on algebraic geometry is Hartshorne [51]. More recent books that discuss algebraic geometry from various angles are Bump [13], Cox, Little, and O'Shea [23], Kunz [65], and Smith et al. [115].

2.1 Affine and Projective Spaces

From now on in this chapter, we assume that k is a perfect field, that is, that every algebraic extension of k is separable. Note that, in particular, the finite field $\mathbb{F}_q$ and its algebraic closure $\overline{\mathbb{F}_q}$ are perfect for any prime power q.

For a given positive integer n, *affine n-space* over k, denoted by $\mathbf{A}^n$, is the Cartesian n-space

$$\mathbf{A}^n := \mathbf{A}^n(\overline{k}) := \{(a_1, \ldots, a_n) : a_i \in \overline{k} \quad \text{for } i = 1, \ldots, n\},$$

where $\overline{k}$ is a fixed algebraic closure of k. For a finite extension K of k, the *K-rational point set* is the subset

$$\mathbf{A}^n(K) := \{(a_1, \ldots, a_n) : a_i \in K \quad \text{for } i = 1, \ldots, n\}.$$

An element $P = (a_1, \ldots, a_n)$ of $\mathbf{A}^n$ is called a *point* and each a_i is called a *coordinate* of P. A point $P \in \mathbf{A}^n(K)$ is called *K-rational*. We also call

P rational if P is k-rational. It is clear that the number of points of $\mathbf{A}^n(\mathbb{F}_{q^m})$ is q^{mn}.

Consider the Galois action of $\mathrm{Gal}(\overline{k}/k)$ on $\mathbf{A}^n$ given by

$$\sigma(P) = (\sigma(a_1), \ldots, \sigma(a_n))$$

for any $\sigma \in \mathrm{Gal}(\overline{k}/k)$ and $P = (a_1, \ldots, a_n) \in \mathbf{A}^n$. It is obvious that for any two automorphisms σ and τ in $\mathrm{Gal}(\overline{k}/k)$ we have $(\sigma\tau)(P) = \sigma(\tau(P))$. In particular, if $k = \mathbb{F}_q$, then we define the Frobenius action by

$$\pi(a_1, \ldots, a_n) = (a_1^q, \ldots, a_n^q).$$

Using the above action, we can characterize the k-rational points by

$$\mathbf{A}^n(k) = \{P \in \mathbf{A}^n : \sigma(P) = P \quad \text{for all } \sigma \in \mathrm{Gal}(\overline{k}/k)\}.$$

If $k = \mathbb{F}_q$, then $\mathbf{A}^n(\mathbb{F}_q) = \{P \in \mathbf{A}^n : \pi(P) = P\}$.

For a point $P \in \mathbf{A}^n$, we call the set

$$\{\sigma(P) : \sigma \in \mathrm{Gal}(\overline{k}/k)\}$$

a *closed point* over k or a *k-closed point* (we just say closed point if there is no confusion). Two points in a closed point over k are called *k-conjugate*.

Lemma 2.1.1. An $\mathbb{F}_q$-closed point is a finite subset of $\mathbf{A}^n(\overline{\mathbb{F}_q})$.

Proof. Let $P = (a_1, \ldots, a_n) \in \mathbf{A}^n(\overline{\mathbb{F}_q})$ be a point. Then for each $i = 1, \ldots, n$, there exists an integer $m_i \geq 1$ such that $a_i \in \mathbb{F}_{q^{m_i}}$. Put $m = \prod_{i=1}^n m_i$. Then by Theorem 1.1.5 (iii), P is an $\mathbb{F}_{q^m}$-rational point. Hence, by Corollary 1.2.7,

$$|\{\sigma(P) : \sigma \in \mathrm{Gal}(\overline{\mathbb{F}_q}/\mathbb{F}_q)\}| = |\{\sigma(P) : \sigma \in \mathrm{Gal}(\mathbb{F}_{q^m}/\mathbb{F}_q)\}|$$
$$< |\mathrm{Gal}(\mathbb{F}_{q^m}/\mathbb{F}_q)| = m.$$

This completes the proof. $\qquad\qquad\qquad\qquad\qquad\qquad\qquad\qquad\square$

We define the *degree* of an $\mathbb{F}_q$-closed point P to be the cardinality of P, denoted by deg(P). The following lemma presents, in particular, some alternative characterizations of deg(P).

Lemma 2.1.2. Let $P = (a_1, \ldots, a_n)$ be an arbitrary point in an $\mathbb{F}_q$-closed point P. Then:

(i) $\mathsf{P} = \{\sigma(P) : \sigma \in \mathrm{Gal}(\overline{\mathbb{F}_q}/\mathbb{F}_q)\}$;

(ii) $\mathsf{P} = \{\pi^j(P) : j = 0, 1, \ldots\}$, where $\pi \in \mathrm{Gal}(\overline{\mathbb{F}_q}/\mathbb{F}_q)$ is the Frobenius;

(iii) the degree of P is the least positive integer m such that $\pi^m(P) = P$;

(iv) the degree of P is the least positive integer m such that $a_i \in \mathbb{F}_{q^m}$ for all $i = 1, \ldots, n$, that is, m satisfies $\mathbb{F}_{q^m} = \mathbb{F}_q(a_1, \ldots, a_n)$.

Proof.

(i) Let $\mathsf{P} = \{\sigma(Q) : \sigma \in \mathrm{Gal}(\overline{\mathbb{F}_q}/\mathbb{F}_q)\}$ for some point Q. Then $P = \tau(Q) \in \mathsf{P}$ for some $\tau \in \mathrm{Gal}(\overline{\mathbb{F}_q}/\mathbb{F}_q)$. Thus,

$$
\begin{aligned}
\mathsf{P} &= \{\sigma(Q) : \sigma \in \mathrm{Gal}(\overline{\mathbb{F}_q}/\mathbb{F}_q)\} \\
&= \{\sigma(\tau^{-1}(P)) : \sigma \in \mathrm{Gal}(\overline{\mathbb{F}_q}/\mathbb{F}_q)\} \\
&= \{(\sigma\tau^{-1})(P) : \sigma \in \mathrm{Gal}(\overline{\mathbb{F}_q}/\mathbb{F}_q)\} \\
&= \{\sigma(P) : \sigma \in \mathrm{Gal}(\overline{\mathbb{F}_q}/\mathbb{F}_q)\}.
\end{aligned}
$$

(ii) There exists an integer $r \geq 1$ such that $\mathbb{F}_{q^r}$ contains all coordinates a_i. For any automorphism $\sigma \in \mathrm{Gal}(\overline{\mathbb{F}_q}/\mathbb{F}_q)$, $\sigma|_{\mathbb{F}_{q^r}}$ is an automorphism in $\mathrm{Gal}(\mathbb{F}_{q^r}/\mathbb{F}_q)$. Since $\pi|_{\mathbb{F}_{q^r}}$ is a generator of $\mathrm{Gal}(\mathbb{F}_{q^r}/\mathbb{F}_q)$ by Theorem 1.1.5(ii), there exists an integer $j \geq 0$ such that $\sigma|_{\mathbb{F}_{q^r}} = \pi^j|_{\mathbb{F}_{q^r}}$. Hence, $\sigma(P) = \sigma|_{\mathbb{F}_{q^r}}(P) = \pi^j|_{\mathbb{F}_{q^r}}(P) = \pi^j(P)$.

(iii) This follows directly from (ii).

(iv) This is shown by similar arguments as in the proof of Proposition 1.3.2. $\qquad\square$

For a point $P = (a_1, \ldots, a_n) \in \mathbf{A}^n$, the field $k(a_1, \ldots, a_n)$ is called the *definition field* of P over k, denoted by $k(P)$.

For a point $P = (a_1, \ldots, a_n) \in \mathbf{A}^n(\overline{\mathbb{F}_q})$ and $\sigma \in \mathrm{Gal}(\overline{\mathbb{F}_q}/\mathbb{F}_q)$, we note that $\mathbb{F}_q(\sigma(a_1), \ldots, \sigma(a_n)) = \sigma(\mathbb{F}_q(a_1, \ldots, a_n))$ is $\mathbb{F}_q$-isomorphic to $\mathbb{F}_q(a_1, \ldots, a_n)$. By the uniqueness of finite fields, these two fields are the same. This means that P and $\sigma(P)$ have the same definition field. Hence, we can also speak of the *definition field* of an $\mathbb{F}_q$-closed point P, denoted by $\mathbb{F}_q(\mathsf{P})$. Thus, Lemma 2.1.2(iv) says that the degree of an $\mathbb{F}_q$-closed point P is equal to $[\mathbb{F}_q(\mathsf{P}) : \mathbb{F}_q]$.

Proposition 2.1.3. For positive integers m and r, an $\mathbb{F}_q$-closed point P of degree m splits into $\gcd(r, m)$ $\mathbb{F}_{q^r}$-closed points of degree $m/\gcd(r, m)$, that is, $\mathsf{P} = \cup_{j=1}^{\gcd(r,m)} \mathsf{P}_j$ with $\mathsf{P}_1, \ldots, \mathsf{P}_{\gcd(r,m)}$ disjoint and each P_j being an $\mathbb{F}_{q^r}$-closed point of degree $m/\gcd(r, m)$.

Proof. Let P be a point in P. Then by Lemma 2.1.2(iii), the degree of the $\mathbb{F}_{q^r}$-closed point containing P is the least positive integer t such that $(\pi^r)^t(P) = P$, where π is the Frobenius of $\overline{\mathbb{F}_q}/\mathbb{F}_q$. Since m is the least positive integer satisfying $\pi^m(P) = P$, we obtain $t = m/\gcd(r, m)$. Thus, P splits into $\gcd(r, m)$ $\mathbb{F}_{q^r}$-closed points.   □

Corollary 2.1.4. For positive integers m and r, an $\mathbb{F}_q$-closed point of degree m splits into m $\mathbb{F}_{q^r}$-closed points of degree 1 if and only if m divides r.

Corollary 2.1.5. An irreducible polynomial of degree m over $\mathbb{F}_q$ is equal to a product of $\gcd(r, m)$ irreducible polynomials of degree $m/\gcd(r, m)$ over $\mathbb{F}_{q^r}$.

Proof. Consider the affine space $\mathbf{A}^1(\overline{\mathbb{F}_q})$. Then it follows from Corollary 1.3.4 that an $\mathbb{F}_q$-closed point of degree m corresponds to the set of roots of an irreducible polynomial of degree m over $\mathbb{F}_q$. The desired result follows from Proposition 2.1.3.   □

Theorem 2.1.6. For a positive integer m, let $I_q(n, m)$ be the number of $\mathbb{F}_q$-closed points of degree m in $\mathbf{A}^n(\overline{\mathbb{F}_q})$. Then

$$I_q(n, m) = \frac{1}{m} \sum_{d \mid m} \mu(d) q^{mn/d},$$

where the sum is over all positive integers d dividing m and μ is the Möbius function on $\mathbb{N}$.

Proof. It follows from Corollary 2.1.4 that $\sum_{d \mid m} d I_q(n, d) = q^{mn}$ for all $m \geq 1$. The Möbius inversion formula (see [72, p. 92]) gives the desired result.   □

Roughly speaking, projective spaces are obtained by adding "points at infinity" to affine spaces. In order to do so, we have to go to affine spaces of the next higher dimension.

Definition 2.1.7. A *projective n-space* over k, denoted by $\mathbf{P}^n(\overline{k})$ or $\mathbf{P}^n$, is the set of equivalence classes of nonzero $(n + 1)$-tuples $(a_0, a_1, \ldots, a_n)$ of elements of $\overline{k}$ under the equivalence relation given by

$$(a_0, a_1, \ldots, a_n) \sim (b_0, b_1, \ldots, b_n)$$

if and only if there exists a nonzero element λ of $\overline{k}$ such that $a_i = \lambda b_i$ for all $i = 0, 1, \ldots, n$. An element of $\mathbf{P}^n$ is called a *point*. We denote by $[a_0, a_1, \ldots, a_n]$ the equivalence class containing the $(n + 1)$-tuple $(a_0, a_1, \ldots, a_n)$. Thus, $[a_0, a_1, \ldots, a_n]$ and $[\lambda a_0, \lambda a_1, \ldots, \lambda a_n]$ stand for the same point if $\lambda \neq 0$. For a point $P = [a_0, a_1, \ldots, a_n]$, each a_i is called a *homogeneous coordinate* of P. For a finite extension K of k, a point in the set

$$\mathbf{P}^n(K) := \{[a_0, a_1, \ldots, a_n] \in \mathbf{P}^n : a_i \in K \quad \text{for } i = 0, 1, \ldots, n\}$$

is called *K-rational*. A k-rational point is also said to be *rational*.

Since a point in $\mathbf{P}^n$ is an equivalence class, the point $[a_0, a_1, \ldots, a_n]$ of $\mathbf{P}^n$ is K-rational if and only if there exists a nonzero element $\lambda \in \overline{k}$ such that $\lambda a_i \in K$ for all $i = 0, 1, \ldots, n$. This is equivalent to saying that

$$(a_0/a_j, a_1/a_j, \ldots, a_{j-1}/a_j, a_{j+1}/a_j, \ldots, a_n/a_j) \in \mathbf{A}^n(K)$$

for any j with $a_j \neq 0$.

Proposition 2.1.8. The number of $\mathbb{F}_q$-rational points in $\mathbf{P}^n(\overline{\mathbb{F}_q})$ is given by

$$\frac{q^{n+1} - 1}{q - 1}.$$

Proof. Since every $\mathbb{F}_q$-rational point contains an $(n+1)$-tuple belonging to $\mathbf{A}^{n+1}(\mathbb{F}_q)$, we can restrict the attention to the set $\mathbf{A}^{n+1}(\mathbb{F}_q)$. It is clear that two nonzero $(n + 1)$-tuples $(a_0, a_1, \ldots, a_n)$ and $(b_0, b_1, \ldots, b_n)$ in $\mathbf{A}^{n+1}(\mathbb{F}_q)$ are in the same equivalence class if and only if there exists a nonzero element $\lambda \in \mathbb{F}_q$ such that $a_i = \lambda b_i$ for $i = 0, 1, \ldots, n$. Thus, there are exactly $(q^{n+1} - 1)/(q - 1)$ distinct classes in $\mathbf{A}^{n+1}(\mathbb{F}_q)$, and so there are exactly $(q^{n+1} - 1)/(q - 1)$ distinct $\mathbb{F}_q$-rational points. $\square$

Consider the Galois action of $\mathrm{Gal}(\overline{k}/k)$ on $\mathbf{P}^n$ given by

$$\sigma([a_0, a_1, \ldots, a_n]) = [\sigma(a_0), \sigma(a_1), \ldots, \sigma(a_n)]$$

for any $\sigma \in \mathrm{Gal}(\overline{k}/k)$ and $[a_0, a_1, \ldots, a_n] \in \mathbf{P}^n$. Since $\sigma([\lambda a_0, \lambda a_1, \ldots, \lambda a_n]) = [\sigma(\lambda)\sigma(a_0), \sigma(\lambda)\sigma(a_1), \ldots, \sigma(\lambda)\sigma(a_n)]$, the above action is well defined. In particular, for $k = \mathbb{F}_q$ we define the Frobenius action by

$$\pi([a_0, a_1, \ldots, a_n]) = [a_0^q, a_1^q, \ldots, a_n^q].$$

Using the above action, we can characterize the k-rational points by

$$\mathbf{P}^n(k) = \{P \in \mathbf{P}^n : \sigma(P) = P \quad \text{for all } \sigma \in \mathrm{Gal}(\overline{k}/k)\}.$$

Similarly, we can define closed points in $\mathbf{P}^n$. For a point $P \in \mathbf{P}^n$, we call the set

$$\{\sigma(P) : \sigma \in \mathrm{Gal}(\overline{k}/k)\}$$

a *closed point* over k or a *k-closed* point (we just say closed point if there is no confusion). Two points in a closed point over k are called *k-conjugate*.

Lemma 2.1.9. An $\mathbb{F}_q$-closed point is a finite subset of $\mathbf{P}^n(\overline{\mathbb{F}_q})$.

Proof. Let $P = [a_0, a_1, \ldots, a_n] \in \mathbf{P}^n(\overline{\mathbb{F}_q})$ be a point. Then for each $i = 0, 1, \ldots, n$, there exists an integer $m_i \geq 1$ such that $a_i \in \mathbb{F}_{q^{m_i}}$. Put $m = \prod_{i=0}^n m_i$. Then by Theorem 1.1.5(iii), P is an $\mathbb{F}_{q^m}$-rational point. Hence, by Corollary 1.2.7,

$$|\{\sigma(P) : \sigma \in \mathrm{Gal}(\overline{\mathbb{F}_q}/\mathbb{F}_q)\}| = |\{\sigma(P) : \sigma \in \mathrm{Gal}(\mathbb{F}_{q^m}/\mathbb{F}_q)\}|$$
$$\leq |\mathrm{Gal}(\mathbb{F}_{q^m}/\mathbb{F}_q)| = m.$$

This completes the proof. □

We define the *degree* of an $\mathbb{F}_q$-closed point P to be the cardinality of P, denoted by deg(P).

For a point $P = [a_0, a_1, \ldots, a_n] \in \mathbf{P}^n$ with $a_i \neq 0$, the *definition field* of P over k, denoted by $k(P)$, is defined by $k(a_0/a_i, a_1/a_i, \ldots, a_n/a_i)$.

Remark 2.1.10.

(i) The definition field of a point is well defined. First of all, if $(b_0, b_1, \ldots, b_n) = (\lambda a_0, \lambda a_1, \ldots, \lambda a_n)$ with $\lambda \neq 0$, we have $b_j/b_i = a_j/a_i$ for all $j = 0, 1, \ldots, n$. Hence, $k(a_0/a_i, a_1/a_i, \ldots, a_n/a_i) = k(b_0/b_i, b_1/b_i, \ldots, b_n/b_i)$. Second, without loss of generality, we assume that $a_0 \neq 0$ and $a_1 \neq 0$. Put $c := a_1/a_0$. Then

$$
\begin{aligned}
k(a_0/a_0, a_1/a_0, \ldots, a_n/a_0) &= k(c(a_0/a_1), c(a_1/a_1), \ldots, c(a_n/a_1)) \\
&\subseteq k(a_0/a_1, a_1/a_1, \ldots, a_n/a_1) \\
&= k(c^{-1}(a_0/a_0), c^{-1}(a_1/a_0), \ldots, c^{-1}(a_n/a_0)) \\
&\subseteq k(a_0/a_0, a_1/a_0, \ldots, a_n/a_0),
\end{aligned}
$$

and so $k(a_0/a_1, a_1/a_1, \ldots, a_n/a_1) = k(a_0/a_0, a_1/a_0, \ldots, a_n/a_0)$.

(ii) If P and Q are two k-conjugate points, then it is clear that for a fixed i, the ith homogeneous coordinate of P is zero if and only if the ith homogeneous coordinate of Q is zero.

(iii) Let P and Q be two points in the same $\mathbb{F}_q$-closed point P. We may assume that $P = [1, a_1, \ldots, a_n]$ and $Q = \sigma(P) = [1, \sigma(a_1), \ldots, \sigma(a_n)]$ for some $\sigma \in \mathrm{Gal}(\overline{\mathbb{F}_q}/\mathbb{F}_q)$. Then $\mathbb{F}_q(P) = \mathbb{F}_q(a_1, \ldots, a_n) = \mathbb{F}_q(\sigma(a_1), \ldots, \sigma(a_n)) = \mathbb{F}_q(Q)$. This means that the definition field for any $\mathbb{F}_q$-conjugate points is the same. Hence, we can also speak of the *definition field* of an $\mathbb{F}_q$-closed point, denoted by $\mathbb{F}_q(\mathsf{P})$.

For each $i = 1, \ldots, n + 1$, we consider the map

$$
\phi_i : \mathbf{A}^n \to \mathbf{P}^n, \quad (a_1, \ldots, a_n) \mapsto [a_1, \ldots, a_{i-1}, 1, a_i, \ldots, a_n]. \tag{2.1}
$$

Then ϕ_i is injective. Thus, $\mathbf{A}^n$ can be embedded into $\mathbf{P}^n$. Moreover, for a point $P \in \mathbf{A}^n$, it is easy to see that $\{\sigma(\phi_i(P)) : \sigma \in \mathrm{Gal}(\overline{k}/k)\} = \{\phi_i(\sigma(P)) : \sigma \in \mathrm{Gal}(\overline{k}/k)\}$. This means that ϕ_i induces a bijective map between the closed points of $\mathbf{A}^n$ and those closed points of $\mathbf{P}^n$ with the ith homogeneous coordinate not equal to zero. Furthermore, the degree of an $\mathbb{F}_q$-closed point P of $\mathbf{A}^n$ is equal to the degree of $\phi_i(\mathsf{P})$.

Lemma 2.1.11. Let $P \in \mathbf{P}^n$ be an arbitrary point in an $\mathbb{F}_q$-closed point P. Then:

(i) $\mathsf{P} = \{\sigma(P) : \sigma \in \mathrm{Gal}(\overline{\mathbb{F}_q}/\mathbb{F}_q)\}$;

(ii) $P = \{\pi^j(P) : j = 0, 1, \ldots\}$, where $\pi \in \mathrm{Gal}(\overline{\mathbb{F}_q}/\mathbb{F}_q)$ is the Frobenius;

(iii) the degree of P is the least positive integer m such that $\pi^m(P) = P$;

(iv) the degree of P is equal to the degree of the extension $\mathbb{F}_q(P)$ over $\mathbb{F}_q$.

Proof. (i), (ii), and (iii) are shown in the same way as the corresponding results in Lemma 2.1.2. As to (iv), assume that the ith homogeneous coordinate of P is not equal to zero. Then $\phi_i^{-1}(P)$ is a closed point of $\mathbf{A}^n$. The degree of P is equal to the degree of $\phi_i^{-1}(P)$, which is equal to $[\mathbb{F}_q(P) : \mathbb{F}_q]$ by Lemma 2.1.2(iv). $\qquad\qquad\square$

The following result is the projective analog of Proposition 2.1.3.

Proposition 2.1.12. For positive integers m and r, an $\mathbb{F}_q$-closed point P of degree m splits into $\gcd(r, m)$ $\mathbb{F}_{q^r}$-closed points of degree $m/\gcd(r, m)$, that is, $P = \bigcup_{j=1}^{\gcd(r,m)} P_j$ with $P_1, \ldots, P_{\gcd(r,m)}$ disjoint and each P_j being an $\mathbb{F}_{q^r}$-closed point of degree $m/\gcd(r, m)$.

Proof. Let P contain the point P. Without loss of generality, we may assume that the first coordinate of P is not zero. Then $P \subset \phi_1(\mathbf{A}^n)$ and the desired result follows from Proposition 2.1.3. $\qquad\qquad\square$

Corollary 2.1.13. For positive integers m and r, an $\mathbb{F}_q$-closed point of degree m splits into m $\mathbb{F}_{q^r}$-closed points of degree 1 if and only if m divides r.

2.2 Algebraic Sets

In analogy with the distinction between affine and projective spaces in Section 2.1, we distinguish between affine and projective algebraic sets. We start with a discussion of affine algebraic sets.

Let $k[X] := k[x_1, \ldots, x_n]$ and $\overline{k}[X] := \overline{k}[x_1, \ldots, x_n]$ be the polynomial rings over k and $\overline{k}$, respectively, in the variables $x_1, \ldots, x_n$. For a polynomial $f \in \overline{k}[X]$, we denote the *zero set* of f by

$$Z(f) = \{P \in \mathbf{A}^n = \mathbf{A}^n(\overline{k}) : f(P) = 0\}.$$

More generally, for a subset S of $\bar{k}[X]$ we define the *zero set* $Z(S)$ of S to be the set of common zeros of all polynomials in S, that is,

$$Z(S) = \{P \in \mathbf{A}^n : f(P) = 0 \quad \text{for all } f \in S\}.$$

If $S = \{f_1, \ldots, f_t\}$ is finite, then we also write $Z(S) = Z(f_1, \ldots, f_t)$.

Definition 2.2.1. An *affine algebraic set* is any set of the form $Z(S)$ for some subset S of $\bar{k}[X]$. The set $Z(S)$ is said to be *defined over k* if S is a subset of $k[X]$.

We denote by V/k an affine algebraic set V defined over k. The set of k-rational points of an affine algebraic set V/k defined over k is

$$V(k) := V \cap \mathbf{A}^n(k),$$

that is, a point $P \in V$ is k-rational if and only if $\sigma(P) = P$ for all $\sigma \in \mathrm{Gal}(\bar{k}/k)$.

For an n-tuple $J = (j_1, \ldots, j_n)$ of nonnegative integers, we abbreviate $x_1^{j_1} \cdots x_n^{j_n}$ by X^J. For a polynomial $f(X) = \sum c_J X^J \in \bar{k}[X]$ with all $c_J \in \bar{k}$ and an automorphism $\sigma \in \mathrm{Gal}(\bar{k}/k)$, we define the action $\sigma(f(X)) = \sum \sigma(c_J) X^J$. Thus, $f(X)$ belongs to $k[X]$ if and only if $\sigma(f) = f$ for all $\sigma \in \mathrm{Gal}(\bar{k}/k)$. This is equivalent to $\pi(f) = f$ for the Frobenius $\pi \in \mathrm{Gal}(\overline{\mathbb{F}_q}/\mathbb{F}_q)$ if $k = \mathbb{F}_q$.

Lemma 2.2.2. For any point $P \in \mathbf{A}^n$, any polynomial $f(X) \in \bar{k}[X]$, and any automorphism $\sigma \in \mathrm{Gal}(\bar{k}/k)$, we have $\sigma(f(P)) = \sigma(f)(\sigma(P))$.

Proof. Let $P = (a_1, \ldots, a_n)$. For $J = (j_1, \ldots, j_n)$, put $P^J = a_1^{j_1} \cdots a_n^{j_n}$. Then it is clear that $\sigma(P^J) = \sigma(P)^J$ for any $\sigma \in \mathrm{Gal}(\bar{k}/k)$. For $f(X) = \sum c_J X^J$, we get

$$\sigma(f(P)) = \sigma\left(\sum c_J P^J\right) = \sum \sigma(c_J) \sigma(P^J)$$

$$= \sum \sigma(c_J) \sigma(P)^J = \sigma(f)(\sigma(P)).$$

This completes the proof. $\qquad\qquad\qquad\qquad\qquad\qquad\qquad\square$

Proposition 2.2.3. Let P be a point in a k-closed point P. If P is a point of an affine algebraic set V/k, then $\mathsf{P} \subseteq V$.

Proof. Let $\sigma(P)$ be an arbitrary point in P. Let $V = Z(S)$ for some $S \subseteq k[X]$. For any polynomial $f \in S$, we have by Lemma 2.2.2,

$$f(\sigma(P)) = \sigma(\sigma^{-1}(f))(\sigma(P)) = \sigma(\sigma^{-1}(f)(P)) = \sigma(f(P)) = \sigma(0) = 0.$$

This means that P is a subset of V.   □

By the above result, it makes sense to speak of a k-*closed point* of an affine algebraic set defined over k.

Example 2.2.4.

(i) Let q be a power of 2. Let $V \subseteq \mathbf{A}^2(\overline{\mathbb{F}_q})$ be the affine algebraic set defined by the single equation

$$y^2 + y - x = 0.$$

By Theorem 1.4.3(ii), for an element $a \in \mathbb{F}_q$ there exists an element $b \in \mathbb{F}_q$ such that $(a, b) \in V$ if and only if a is in the kernel of the trace map $\mathrm{Tr}_{\mathbb{F}_q/\mathbb{F}_2}$. Furthermore, if (a, b) is in V, then $(a, b + 1)$ is also in V. It is clear that the number of $\mathbb{F}_q$-rational points of V is equal to q.

(ii) Consider the affine algebraic set $V \subseteq \mathbf{A}^2(\overline{\mathbb{F}_2})$ defined by the single equation

$$y^2 + y - x^3 - x = 0.$$

Then V has four $\mathbb{F}_2$-rational points $(0, 0), (0, 1), (1, 0), (1, 1)$. Moreover, these four points are also all $\mathbb{F}_4$-rational points of V. By Lemma 2.1.2(iv), this shows that V has no $\mathbb{F}_2$-closed points of degree 2.

Example 2.2.5. Let $P = (a_1, \ldots, a_n)$ be a point in $\mathbf{A}^n(\overline{k})$. Then it is obvious that $\{P\} = Z(x_1 - a_1, \ldots, x_n - a_n)$. It can be shown that this gives a one-to-one correspondence between the maximal ideals of $\overline{k}[X]$ and the points of $\mathbf{A}^n(\overline{k})$.

The following properties are easy to verify.

Lemma 2.2.6.

(i) Let S be a nonempty subset of $\bar{k}[X]$. If I is the ideal of $\bar{k}[X]$ generated by S, then $Z(I) = Z(S)$.

(ii) For any two subsets $T \subseteq S$ of $\bar{k}[X]$, we have $Z(S) \subseteq Z(T)$.

(iii) If $\{S_j\}$ is any nonempty collection of subsets of $\bar{k}[X]$, then $Z(\cup S_j) = \cap Z(S_j)$; so the intersection of any nonempty collection of affine algebraic sets is an affine algebraic set.

(iv) $Z(fg) = Z(f) \cup Z(g)$ for any polynomials $f, g \in \bar{k}[X]$ and $Z(S) \cup Z(T) = Z(\{fg : f \in S,\ g \in T\})$ for any $S, T \subseteq \bar{k}[X]$; so any finite union of affine algebraic sets is an affine algebraic set.

(v) $Z(0) = \mathbf{A}^n$ and $Z(1) = \varnothing$.

For an affine algebraic set V/k defined over k, we put

$$I(V) = \{f \in \bar{k}[X] : f(P) = 0 \quad \text{for all } P \in V\}$$

and

$$I(V/k) = I(V) \cap k[X] = \{f \in k[X] : f(P) = 0 \quad \text{for all } P \in V\}.$$

Then it is easy to prove that $I(V)$ (respectively $I(V/k)$) is an ideal of $\bar{k}[X]$ (respectively $k[X]$).

Example 2.2.7. Let $n = 1$ and $V \subseteq \mathbf{A}^1$ be an affine algebraic set. Then $I(V)$ is a principal ideal (g) of $\bar{k}[X]$ for some $g \in \bar{k}[X]$. If $I(V) = \{0\}$, then $V = \mathbf{A}^1$, and if $I(V) = \bar{k}[X]$, then $V = \varnothing$. Otherwise, g has a positive degree and roots $b_1, \ldots, b_d \in \bar{k} = \mathbf{A}^1$. Then $V = \{b_1, \ldots, b_d\}$. Thus, the affine algebraic sets in $\mathbf{A}^1$ are $\mathbf{A}^1$ itself and the finite subsets of $\mathbf{A}^1$. If V is defined over k, then $I(V/k)$ is a principal ideal of $k[X]$. In the case where this ideal is nontrivial, we have $I(V/k) = (g)$ with $g \in k[X]$ of positive degree. For each monic irreducible factor of g in $k[X]$, the roots of this factor form a k-closed point of V.

In the proof of the following proposition, we use two fundamental results on polynomial rings, namely the Hilbert basis theorem and the Hilbert Nullstellensatz, which can be found, for example, in the books [23] and [29].

Proposition 2.2.8. Let V be an affine algebraic set defined over k. Then:

(i) there exists a finite subset T of $k[X]$ such that V is the zero set $Z(T)$ of T;

(ii) $Z(I(V)) = V$;

(iii) if $V = Z(I)$ for some ideal I of $\bar{k}[X]$, then $I(V)$ is equal to the radical of I, that is,

$$I(V) = \{ f \in \bar{k}[X] : f^r \in I \quad \text{for some integer } r \geq 1 \}.$$

Proof.

(i) We have $V = Z(S)$ for some nonempty subset S of $k[X]$. If I is the ideal of $k[X]$ generated by S, then $V = Z(I)$. By the Hilbert basis theorem, I is generated by a finite subset T of $k[X]$, and so $V = Z(T)$.

(ii) Since $V = Z(S)$, it is trivial that $S \subseteq I(V)$, and so we get $Z(I(V)) \subseteq Z(S) = V$ by Lemma 2.2.6(ii). It follows from the definitions that $V \subseteq Z(I(V))$.

(iii) Again by the Hilbert basis theorem, I is generated by a finite subset T of $\bar{k}[X]$. If $f \in I(V) = I(Z(T))$, then the Hilbert Nullstellensatz shows that $f^r \in I$ for some integer $r \geq 1$. Conversely, if $f \in \bar{k}[X]$ is such that $f^r \in I$ for some integer $r \geq 1$, then $f^r(P) = 0$ for all $P \in V$; hence, $f(P) = 0$ for all $P \in V$, and so $f \in I(V)$. $\square$

Remark 2.2.9. If $V = Z(I)$ for some prime ideal I of $\bar{k}[X]$, then $I(V) = I$ by Proposition 2.2.8(iii), and so $I(V)$ is also a prime ideal of $\bar{k}[X]$.

Now we turn to projective algebraic sets. In the remainder of this section, we denote $(x_0, x_1, \ldots, x_n)$ by X.

Definition 2.2.10. A polynomial $f \in \bar{k}[X] = \bar{k}[x_0, x_1, \ldots, x_n]$ is called *homogeneous* of degree d if

$$f(\lambda x_0, \lambda x_1, \ldots, \lambda x_n) = \lambda^d f(x_0, x_1, \ldots, x_n)$$

for all $\lambda \in \bar{k}$. An ideal of $\bar{k}[X]$ is *homogeneous* if it is generated by homogeneous polynomials.

Definition 2.2.11. A *projective algebraic set* is any set V of the form

$$Z_h(S) = \{P \in \mathbf{P}^n : f(P) = 0 \text{ for all homogeneous } f \in S\}$$

for some subset S of $\bar{k}[X]$, and V is said to be *defined over k* if S is a subset of $k[X]$.

We use a similar convention as before, namely, we write $Z_h(S) = Z_h(f_1, \ldots, f_t)$ if $S = \{f_1, \ldots, f_t\}$ is a finite subset of $\bar{k}[X]$.

We denote by V/k a projective algebraic set V defined over k. The set of k-rational points of a projective algebraic set V/k defined over k is

$$V(k) = V \cap \mathbf{P}^n(k),$$

that is, a point $P \in V$ is k-rational if and only if $\sigma(P) = P$ for all $\sigma \in \mathrm{Gal}(\bar{k}/k)$.

Example 2.2.12. A *line* in $\mathbf{P}^2(k)$ is a projective algebraic set given by a linear equation

$$ax_0 + bx_1 + cx_2 = 0,$$

where $a, b, c \in k$ are not all zero. More generally, a *hyperplane* in $\mathbf{P}^n(k)$ is given by an equation

$$a_0x_0 + a_1x_1 + \cdots + a_nx_n = 0,$$

where $a_0, a_1, \ldots, a_n \in k$ are not all zero.

For projective algebraic sets, we have the following properties, which are easy to prove.

Lemma 2.2.13.

(i) Let S be a nonempty set of homogeneous polynomials from $\bar{k}[X]$. If I is the ideal of $\bar{k}[X]$ generated by S, then $Z_h(I) = Z_h(S)$.
(ii) For any two subsets $T \subseteq S$ of $\bar{k}[X]$, we have $Z_h(S) \subseteq Z_h(T)$.
(iii) If $\{S_j\}$ is any nonempty collection of sets of homogeneous polynomials from $\bar{k}[X]$, then $Z_h(\cup S_j) = \cap Z_h(S_j)$; so the intersection of

any nonempty collection of projective algebraic sets is a projective algebraic set.

(iv) $Z_h(fg) = Z_h(f) \cup Z_h(g)$ for any homogeneous polynomials $f, g \in \overline{k}[X]$ and $Z_h(S) \cup Z_h(T) = Z_h(\{fg : f \in S, \ g \in T\})$ for any $S, T \subseteq \overline{k}[X]$; so any finite union of projective algebraic sets is a projective algebraic set.

(v) $Z_h(0) = \mathbf{P}^n$ and $Z_h(1) = \varnothing$.

For any point $P \in \mathbf{P}^n$, any homogeneous polynomial $f(X) \in \overline{k}[X]$, and any automorphism $\sigma \in \mathrm{Gal}(\overline{k}/k)$, we have

$$\sigma(f(P)) = \sigma(f)(\sigma(P)).$$

From this identity we see that a k-closed point P is a subset of V/k as long as one of the points in P belongs to V. Thus, it makes sense to speak of a *k-closed point* of a projective algebraic set defined over k.

For a projective algebraic set V, we let $I(V)$ be the ideal of $\overline{k}[X]$ generated by the set

$$\{f \in \overline{k}[X] : f \text{ is homogeneous and } f(P) = 0 \quad \text{for all } P \in V\}.$$

Note that the ideal $I(V)$ is homogeneous in the sense of Definition 2.2.10.

We prove a lemma on homogeneous ideals, which will be used in the next section. We observe that, by collecting the terms of the same degree, any polynomial in $\overline{k}[X]$ can be written as a sum of finitely many homogeneous polynomials.

Lemma 2.2.14.

(i) An ideal I of $\overline{k}[X]$ is homogeneous if and only if the following condition is satisfied: for every polynomial $f = \sum_d f^{[d]} \in I$ with $f^{[d]}$ being homogeneous of degree d, we also have $f^{[d]} \in I$ for all d.

(ii) A homogeneous proper ideal I of $\overline{k}[X]$ is a prime ideal if and only if the following condition is satisfied: for any two homogeneous polynomials $f, g \in \overline{k}[X]$, if fg belongs to I, then one of f and g must belong to I.

Proof.

(i) Let I be homogeneous and let $f = \sum_{d=0}^{m} f^{[d]}$ be an element of I. By induction, it suffices to show that $f^{[m]}$ belongs to I. Since

I is homogeneous, we can write $f = \sum_{j=1}^{r} h_j f_j$ with $f_j \in \bar{k}[X]$ homogeneous and $h_j \in \bar{k}[X]$ for $1 \le j \le r$. Then $f^{[m]} = \sum_{j=1}^{r} h_j^{[m-\deg(f_j)]} f_j$ is an element of I. Conversely, suppose that the given condition is satisfied. By the Hilbert basis theorem, I is generated by finitely many polynomials $g_1, \ldots, g_t \in \bar{k}[X]$, but they are not necessarily homogeneous. However, if we write each g_l, $1 \le l \le t$, as a sum of homogeneous polynomials $g_l^{[d]}$, $0 \le d \le \deg(g_l)$, then I is generated by the polynomials $g_l^{[d]} \in I$, $0 \le d \le \deg(g_l)$, $1 \le l \le t$, and so I is a homogeneous ideal.

(ii) The necessity is clear. To prove the sufficiency, let $u, v \in \bar{k}[X]$ be two polynomials satisfying $uv \in I$. Suppose that neither u nor v is in I. Put $u = \sum_{d=0}^{m} u^{[d]}$ and $v = \sum_{d=0}^{s} v^{[d]}$. Without loss of generality, we may assume that neither $u^{[m]}$ nor $v^{[s]}$ is in I (otherwise we can consider $u - u^{[m]}$ or $v - v^{[s]}$ instead). By (i), $u^{[m]} v^{[s]} = (uv)^{[m+s]}$ is an element of I. Hence, $u^{[m]} \in I$ or $v^{[s]} \in I$ by the given condition. This is a contradiction. $\qquad\square$

2.3 Varieties

By combining Lemmas 2.2.6 and 2.2.13, we obtain the following proposition.

Proposition 2.3.1. Fix the affine (respectively projective) n-space $\mathbf{A}^n$ (respectively $\mathbf{P}^n$) and consider algebraic subsets of $\mathbf{A}^n$ (respectively $\mathbf{P}^n$). The union of finitely many affine (respectively projective) algebraic sets is an affine (respectively projective) algebraic set. The intersection of any family of affine (respectively projective) algebraic sets is an affine (respectively projective) algebraic set. The empty set and the whole space $\mathbf{A}^n$ (respectively $\mathbf{P}^n$) are affine (respectively projective) algebraic sets.

Remark 2.3.2. It is easy to verify that the union of two algebraic sets V_1 and V_2 is defined over k whenever V_1 and V_2 are defined over k; and the intersection of a family $\{V_j\}$ of algebraic sets is defined over k provided that every algebraic set V_j is defined over k.

The above proposition allows us to define important topologies on $\mathbf{A}^n$ and $\mathbf{P}^n$. The reader may refer to Appendix A.1 for basic results on topology.

Definition 2.3.3. The *Zariski topology* on $\mathbf{A}^n$ (respectively $\mathbf{P}^n$) is defined by taking the open sets to be the complements of the affine (respectively projective) algebraic sets.

The following example studies the relationship between the Zariski topology on $\mathbf{A}^n$ and that on $\mathbf{P}^n$.

Example 2.3.4. For $i = 0, 1, \ldots, n$, let H_i be the closed subset of $\mathbf{P}^n$ defined by $x_i = 0$ and put $U_i = \mathbf{P}^n \setminus H_i$. Then U_i is an open subset of $\mathbf{P}^n$ and $\mathbf{P}^n$ is covered by $\{U_i\}_{i=0}^n$. Define the map

$$\theta_i : U_i \to \mathbf{A}^n, \quad [a_0, a_1, \ldots, a_n] \mapsto \left(\frac{a_0}{a_i}, \ldots, \frac{a_{i-1}}{a_i}, \frac{a_{i+1}}{a_i}, \ldots, \frac{a_n}{a_i} \right). \quad (2.2)$$

It is clear that θ_i is well defined and bijective. The inverse of θ_i is the map ϕ_{i+1} defined in (2.1). Via θ_i, the space $\mathbf{A}^n$ can be viewed as an open subset of $\mathbf{P}^n$. We claim that the map θ_i is a homeomorphism from U_i (with its induced Zariski topology from $\mathbf{P}^n$) onto $\mathbf{A}^n$ (with the Zariski topology). It suffices to check this for $i = 0$. Indeed, let $V = Z_h(S) \subseteq U_0$ be an algebraic set of $\mathbf{P}^n$. Then it is easy to verify that

$$\theta_0(V) = Z(\{f(1, y_1, \ldots, y_n) : \ f(x_0, x_1, \ldots, x_n) \in S \text{ homogeneous}\}).$$

Conversely, let $W = Z(T)$ be a closed subset of $\mathbf{A}^n$. Then it is easy to show that

$$\theta_0^{-1}(W) = U_0 \cap Z_h \left(\left\{ x_0^{\deg(f)} f(x_1/x_0, \ldots, x_n/x_0) : \ f \in T \right\} \right).$$

Hence, both θ_0 and θ_0^{-1} are continuous. Our claim follows. Moreover, one can easily see that $\theta_i(V)$ is defined over k if $V \subseteq U_i$ is defined over k.

Example 2.3.5. Let $n = 1$. Then according to Example 2.2.7, the closed subsets of $\mathbf{A}^1$ are $\mathbf{A}^1$ itself and the finite subsets of $\mathbf{A}^1$. Furthermore, $\mathbf{P}^1 = U_0 \cup H_0$ in the notation of Example 2.3.4. In fact, H_0 consists only of the point $[0, 1]$. Since U_0 (in the induced Zariski topology) is homeomorphic to $\mathbf{A}^1$ by Example 2.3.4, it follows that the closed subsets of U_0 are U_0 itself and the finite subsets of U_0. But U_0 is an open subset of $\mathbf{P}^1$, and so the closed subsets of $\mathbf{P}^1$ are $\mathbf{P}^1$ itself and the finite subsets of $\mathbf{P}^1$.

A nonempty topological space is called *irreducible* if it is not equal to the union of any two proper closed subsets. We refer to Appendix A.1 for basic results on irreducible spaces.

Example 2.3.6. The affine space $\mathbf{A}^1$ is irreducible as every proper closed subset of $\mathbf{A}^1$ is finite, while $\mathbf{A}^1$ is infinite (compare with Example 2.3.5). The projective space $\mathbf{P}^1$ is also irreducible as every proper closed subset of $\mathbf{P}^1$ is finite, while $\mathbf{P}^1$ is infinite (compare again with Example 2.3.5).

Definition 2.3.7. A nonempty affine (respectively projective) algebraic set contained in $\mathbf{A}^n$ (respectively $\mathbf{P}^n$) is called an *affine* (respectively *projective*) *variety* if it is irreducible in the Zariski topology of $\mathbf{A}^n$ (respectively $\mathbf{P}^n$).

Affine and projective varieties can be characterized as follows in terms of the ideal $I(V)$ introduced in Section 2.2.

Theorem 2.3.8. An affine (respectively projective) algebraic set V is an affine (respectively projective) variety if and only if $I(V)$ is a prime ideal of $\overline{k}[X]$.

Proof. We first prove the result for the affine case. Assume that $I(V)$ is a prime ideal of $\overline{k}[X]$. Then $I(V) \neq \overline{k}[X]$, and so V is nonempty. Suppose that V is reducible, that is, there exist two proper closed subsets V_1 and V_2 such that $V = V_1 \cup V_2$. By Proposition 2.2.8(ii), we can choose $f_j \in I(V_j) \setminus I(V)$ for $j = 1, 2$. Then $f_1 f_2 \in I(V)$ by the definition of $I(V)$. This means that $I(V)$ is not a prime ideal. This contradiction implies that V is an affine variety.

Conversely, let V be an affine variety. Then $I(V) \neq \overline{k}[X]$ since V is nonempty. Let $fg \in I(V)$ for some $f, g \in \overline{k}[X]$. Then $V \subseteq Z(fg) = Z(f) \cup Z(g)$ by Lemma 2.2.6(iv) and $V = (Z(f) \cap V) \cup (Z(g) \cap V)$. This forces $Z(f) \cap V = V$ or $Z(g) \cap V = V$, that is, $f \in I(V)$ or $g \in I(V)$. Hence, $I(V)$ is a prime ideal of $\overline{k}[X]$.

We now consider the projective case. If $I(V)$ is a prime ideal of $\overline{k}[X]$, then by using the same arguments as above and the analog of Proposition 2.2.8(ii) for projective algebraic sets, we see that V is irreducible.

If V is a projective variety, then $I(V) \neq \bar{k}[X]$ since $V \neq \varnothing$. Let $f, g \in \bar{k}[X]$ be two homogeneous polynomials satisfying $fg \in I(V)$. Then in the same way as above we can show that $f \in I(V)$ or $g \in I(V)$. It follows from Lemma 2.2.14(ii) that $I(V)$ is a prime ideal of $\bar{k}[X]$. $\square$

Remark 2.3.9. For an affine (respectively projective) algebraic set V/k, we also say that V/k is *absolutely irreducible* if V is irreducible, that is, if V is an affine (respectively projective) variety. Note that it is not enough to check that $I(V/k)$ is a prime ideal of $k[X]$. For example, $V = Z(x^2 + y^2)$ is not irreducible in $\mathbf{A}^2(\overline{\mathbb{F}_3})$, but $I(V/\mathbb{F}_3) = (x^2 + y^2)$ is a prime ideal of $\mathbb{F}_3[x, y]$. On the other hand, if V/k is (absolutely) irreducible, then it is easy to verify that $I(V/k)$ is a prime ideal of $k[X]$.

Example 2.3.10. Let $f(X) \in k[X]$ be an *absolutely irreducible* polynomial (i.e., $f(X)$ is irreducible over $\bar{k}$). Then the ideal (f) is a prime ideal of $\bar{k}[X]$. Thus, $V = Z(f) \subset \mathbf{A}^n$ is an affine variety by Theorem 2.3.8 and Remark 2.2.9, which is called a *plane curve* if $n = 2$, a *surface* if $n = 3$, and, in general, a *hypersurface*.

Example 2.3.11. The affine n-space $\mathbf{A}^n$ (respectively the projective n-space $\mathbf{P}^n$) is an affine (respectively projective) variety as $I(\mathbf{A}^n) = I(\mathbf{P}^n) = (0)$ is a prime ideal of $\bar{k}[X]$.

The following proposition shows that one can easily achieve the transition from affine to projective varieties and vice versa.

Proposition 2.3.12. Let us identify the open subset U_i of $\mathbf{P}^n$ with $\mathbf{A}^n$ through the homeomorphism θ_i defined by (2.2).

(i) Let $W \subseteq \mathbf{A}^n$ be an affine variety. Then the closure $\overline{W}$ in $\mathbf{P}^n$ is a projective variety and $W = \overline{W} \cap \mathbf{A}^n$.

(ii) Let $V \subseteq \mathbf{P}^n$ be a projective variety. Then $V \cap \mathbf{A}^n$ is either empty or an affine variety with $\overline{V \cap \mathbf{A}^n} = V$.

(iii) If an affine (respectively projective) variety V is defined over k, then $\overline{V}$ (respectively $V \cap \mathbf{A}^n$) is also defined over k.

Proof. By Theorem A.1.9(ii), the closure of an affine variety is irreducible in $\mathbf{P}^n$. Hence $\overline{W}$ is a projective variety. By Theorem A.1.9(iii), if $V \cap \mathbf{A}^n \neq \varnothing$, then it is an irreducible closed subset of $\mathbf{A}^n$, and so an affine variety. The remaining results follow immediately from Example 2.3.4. $\qquad\square$

We now associate with each nonempty affine or projective algebraic set its dimension. The reader may have an intuitive notion of dimension, but the formal definition is as follows.

Definition 2.3.13. The *dimension* $\dim(V)$ of a nonempty affine or projective algebraic set V is defined to be its dimension as a topological space in the induced Zariski topology (see Definition A.1.10 for the dimension of a topological space).

The concept of dimension affords a crude classification of varieties. A variety of dimension 1 is called a *curve*.

Example 2.3.14. By Example 2.3.6, the only irreducible closed subsets of $\mathbf{A}^1$ are $\mathbf{A}^1$ and singletons $\{a\}$ for some $a \in \overline{k}$. Thus, the dimension of $\mathbf{A}^1$ is 1 by definition. Similarly, the dimension of $\mathbf{P}^1$ is 1.

Definition 2.3.15. Let V/k be an affine algebraic set. The *coordinate ring* of V is defined by

$$\overline{k}[V] = \overline{k}[X]/I(V).$$

We also put $k[V] = k[X]/I(V/k)$.

Remark 2.3.16. If V/k is an affine variety, then the coordinate ring $\overline{k}[V]$ (respectively $k[V]$) is an integral domain since $I(V)$ (respectively $I(V/k)$) is a prime ideal of $\overline{k}[X]$ (respectively $k[X]$).

Example 2.3.17. Let V be an affine algebraic set and let $\{P_1, \ldots, P_r\}$ be a finite subset of V. By Example 2.2.5 we know that every single point is an affine algebraic set. Thus, any finite set is an affine algebraic set. From Proposition 2.2.8(ii) we infer that the ideal $I(\{P_1, \ldots, P_{j-1}, P_{j+1}, \ldots, P_r\})$ is strictly bigger than the ideal $I(\{P_1, \ldots, P_r\})$. Choose an element f_j from $I(\{P_1, \ldots, P_{j-1}, P_{j+1}, \ldots, P_r\}) \setminus I(\{P_1, \ldots, P_r\})$ for $j = 1, \ldots, r$.

Then it is easy to verify that $f_1, \ldots, f_r$ are linearly independent in the coordinate ring $\bar{k}[V]$ which can be viewed as a vector space over $\bar{k}$. Hence, $r \leq \dim_{\bar{k}}(\bar{k}[V])$. This implies that V is a finite set if $\bar{k}[V]$ has a finite $\bar{k}$-dimension.

Conversely, suppose that $V = \{P_1, \ldots, P_r\} \subset \mathbf{A}^n$ is a finite set. Let $P_j = (a_{1j}, \ldots, a_{nj})$ for $1 \leq j \leq r$ and put $g_i = \prod_{j=1}^{r}(x_i - a_{ij})$ for $1 \leq i \leq n$. Then $g_i \in I(V)$. Thus, in the coordinate ring $\bar{k}[V]$ the monomial x_i^r is a linear combination of $1, x_i, \ldots, x_i^{r-1}$. This means that the set $\{\prod_{i=1}^{n} x_i^{e_i} : 0 \leq e_i \leq r - 1\}$ generates the whole vector space $\bar{k}[V]$. This implies that $\bar{k}[V]$ has a finite $\bar{k}$-dimension.

The following theorem provides an algebraic characterization of the dimension of an affine algebraic set. We refer to Appendix A.2 for background on the Krull dimension of rings.

Theorem 2.3.18. The dimension of a nonempty affine algebraic set V is equal to the Krull dimension of its coordinate ring $\bar{k}[V]$.

Proof. By Theorem 2.3.8, the irreducible closed subsets of V correspond to prime ideals of $\bar{k}[X]$ containing $I(V)$. These, in turn, correspond to prime ideals of $\bar{k}[V]$. Hence, $\dim(V)$ is the length of the longest chain of prime ideals of $\bar{k}[V]$, which is the Krull dimension of $\bar{k}[V]$ by definition. $\square$

Corollary 2.3.19. For every $n \geq 1$, we have $\dim(\mathbf{A}^n) = \dim(\mathbf{P}^n) = n$.

Proof. Since $I(\mathbf{A}^n) = (0)$, the coordinate ring $\bar{k}[\mathbf{A}^n]$ is the polynomial ring $\bar{k}[X] = \bar{k}[x_1, \ldots, x_n]$. By Theorem 2.3.18 we have $\dim(\mathbf{A}^n) = \dim(\bar{k}[X])$, and by Theorem A.2.2(i) the latter Krull dimension is equal to the transcendence degree of $\bar{k}(X)$ over $\bar{k}$, which is n.

As $\mathbf{P}^n$ is covered by $\{U_i\}_{i=0}^{n}$ (see Example 2.3.4), we have

$$\dim(\mathbf{P}^n) = \max_{0 \leq i \leq n} \dim(U_i) = \dim(\mathbf{A}^n) = n$$

by Theorem A.1.11(ii) and the fact that each U_i is homeomorphic to $\mathbf{A}^n$ (see again Example 2.3.4). $\square$

2.4 Function Fields of Varieties

In the following definition, we generalize the concepts of affine and projective varieties. As in Section 2.3, we use the Zariski topology on $\mathbf{A}^n$ and $\mathbf{P}^n$.

Definition 2.4.1. A nonempty intersection of an affine (respectively projective) variety in $\mathbf{A}^n$ (respectively $\mathbf{P}^n$) with an open set in $\mathbf{A}^n$ (respectively $\mathbf{P}^n$) is called a *quasi-affine* (respectively *quasi-projective*) *variety*.

It follows from Theorem A.1.9(i) that any quasi-affine variety and any quasi-projective variety is an irreducible topological space in its induced Zariski topology. We now consider special types of $\bar{k}$-valued functions on quasi-affine and quasi-projective varieties.

Definition 2.4.2.

(i) For a point P of a quasi-affine variety V, a $\bar{k}$-valued function f on V is called *regular* at P if there exists a neighborhood N of P such that $f = a/b$ on N for two polynomials $a, b \in \bar{k}[X]$, where $b(Q) \neq 0$ for all $Q \in N$. The function f is said to be *regular* on a nonempty open subset U of V if it is regular at every point of U.

(ii) Let $V \subseteq \mathbf{P}^n$ be a quasi-projective variety. A point $P \in V$ belongs to $V \cap U_i = V \cap \mathbf{A}^n$ (we identify U_i with $\mathbf{A}^n$) for some i. A $\bar{k}$-valued function f on V is said to be *regular* at P if f is regular at P when restricted to the quasi-affine variety $V \cap \mathbf{A}^n$. The function f is said to be *regular* on a nonempty open subset U of V if it is regular at every point of U.

Example 2.4.3. Consider the affine variety $Z(y^2 + y - x^5) \subset \mathbf{A}^2(\overline{\mathbb{F}_2})$. Then the function x/y^2 is regular at all points except $(0, 0)$.

Let V be a quasi-affine or quasi-projective variety. Consider the set of pairs

$$\mathcal{E} := \{(U, f) : \varnothing \neq U \subseteq V \text{ is open and } f \text{ is a regular function on } U\}.$$

Define a relation $\sim$ on $\mathcal{E}$ as follows:

$$(U, f) \sim (W, g) \Leftrightarrow f|_Y = g|_Y$$

for some nonempty open subset Y of $U \cap W$. Note that $U \cap W \neq \varnothing$ by Theorem A.1.9(iv). The following proposition can be shown in a straight-forward manner.

Proposition 2.4.4. The relation $\sim$ defined above is an equivalence relation.

For two equivalence classes $\overline{(U, f)}$ and $\overline{(W, g)}$, define the sum and the product by

$$\overline{(U, f)} + \overline{(W, g)} = \overline{(U \cap W, f + g)}, \qquad \overline{(U, f)} \times \overline{(W, g)} = \overline{(U \cap W, fg)}.$$

Definition 2.4.5. Let V be a quasi-affine or quasi-projective variety and let P be a point of V. Define the *local ring* at P, denoted by $\mathcal{O}_P(V)$, to be the ring of all equivalence classes $\overline{(U, f)}$ with $P \in U$.

Remark 2.4.6.

(i) Indeed, it is easy to verify that $\mathcal{O}_P(V)$ is a local ring in the sense of Definition A.3.1(i) for a quasi-affine or quasi-projective variety V under the addition and multiplication of equivalence classes defined above. The maximal ideal of $\mathcal{O}_P(V)$ is

$$\mathcal{M}_P(V) := \{\overline{(U, f)} \in \mathcal{O}_P(V) : f(P) = 0\}.$$

(ii) Note that for a quasi-projective variety V and a point P of V, the ring $\mathcal{O}_P(V)$ is the same as $\mathcal{O}_P(V \cap \theta_i^{-1}(\mathbf{A}^n))$ if P belongs to $\theta_i^{-1}(\mathbf{A}^n)$. Thus, we can assume that V is a quasi-affine variety when we talk about a local ring $\mathcal{O}_P(V)$.

Definition 2.4.7. For a nonempty open subset U of a quasi-affine or quasi-projective variety V, we write

$$\mathcal{O}_U(V) := \bigcap_{P \in U} \mathcal{O}_P(V).$$

If there is no danger of confusion, we replace $\mathcal{O}_P(V)$ by $\mathcal{O}_P$ and $\mathcal{O}_U(V)$ by $\mathcal{O}_U$.

Definition 2.4.8. Let V be a quasi-affine or quasi-projective variety. The *function field* of V, denoted by $\bar{k}(V)$, consists of the equivalence classes $\overline{(U, f)}$ defined above, where U is a nonempty open subset of V and $\overline{(U, f)}$ is an element of $\mathcal{O}_U(V)$.

Proposition 2.4.9. The function field $\bar{k}(V)$ of a quasi-affine or quasi-projective variety V is indeed a field.

Proof. For two elements $\overline{(U, f)}$ and $\overline{(W, g)}$ of $\bar{k}(V)$, the intersection $U \cap W$ is nonempty by Theorem A.1.9(iv). Thus, we can carry out the addition and multiplication for elements of $\bar{k}(V)$. It is easy to verify that $\bar{k}(V)$ forms a ring. Let $\overline{(U, f)}$ belong to $\bar{k}(V)$ with $f \neq 0$, that is, the zero set $Z(f)$ does not contain U. Thus, the set $Y := U \setminus (U \cap Z(f))$ is a nonempty open subset of U. Hence, $\overline{(Y, 1/f)}$ is the inverse of $\overline{(U, f)}$. $\qquad\square$

Lemma 2.4.10. Let V be a quasi-affine or quasi-projective variety and let U be a nonempty open subset of V. Then the function field of U is the same as the function field of V.

Proof. It is clear that $\bar{k}(U)$ is a subfield of $\bar{k}(V)$. Now let $\overline{(W, f)}$ be an element of $\bar{k}(V)$. Then by Theorem A.1.9(iv) the intersection $W \cap U$ is a nonempty subset of U. Hence, $\overline{(W, f)}$ and $\overline{(W \cap U, f)}$ are in the same equivalence class and this equivalence class is an element of $\bar{k}(U)$. $\qquad\square$

Remark 2.4.11. Note that for a quasi-affine or quasi-projective variety V, a point $P \in V$, and an open subset U of V containing P, we have natural inclusions

$$\mathcal{O}_V \subseteq \mathcal{O}_U \subseteq \mathcal{O}_P \subseteq \bar{k}(V).$$

Remark 2.4.12. Let $V \subseteq \mathbf{A}^n$ be an affine variety. Then by Remark 2.3.16, the coordinate ring $\bar{k}[V]$ is an integral domain. For two polynomials $g, h \in \bar{k}[V]$ with $h \notin I(V)$, we consider $f = g/h$. The set $U = (\mathbf{A}^n \setminus Z(h)) \cap V$ is a nonempty open set in V and it is clear that f is regular on U and, hence, $\overline{(U, f)} \in \bar{k}(V)$. This implies that the quotient field of $\bar{k}[V]$ can be viewed as a subfield of the function field $\bar{k}(V)$. If V is defined over k, then the quotient field of $k[V]$ is also a subfield of $\bar{k}(V)$.

Theorem 2.4.13. Let $V \subseteq \mathbf{A}^n$ be an affine variety with coordinate ring $\overline{k}[V]$. Then:

(i) $\mathcal{O}_V \cong \overline{k}[V]$;

(ii) the mapping $P \in V \mapsto M_P := \{f \in \overline{k}[V] : f(P) = 0\}$ yields a one-to-one correspondence between the points of V and the maximal ideals of $\overline{k}[V]$;

(iii) for each $P \in V$, the ring $\mathcal{O}_P$ is isomorphic to the localization of $\overline{k}[V]$ at M_P, that is,

$$\mathcal{O}_P \cong \overline{k}[V]_{M_P} := \left\{ \frac{f}{g} : f \in \overline{k}[V], \ g \in \overline{k}[V] \setminus M_P \right\},$$

and $\dim(\mathcal{O}_P) = \dim(V)$;

(iv) $\overline{k}(V)$ is isomorphic to the quotient field of $\overline{k}[V]$ and $\dim(V)$ is equal to the transcendence degree of $\overline{k}(V)$ over $\overline{k}$.

Proof.

(ii) By Example 2.2.5 there is a one-to-one correspondence between the points P of V and the maximal ideals of $\overline{k}[X]$ containing $I(V)$. Passing to the factor ring $\overline{k}[V]$, this means that there is a one-to-one correspondence between the points P of V and the maximal ideals of $\overline{k}[V]$.

(iii) The map $\omega : f \in \overline{k}[V] \mapsto f \in \mathcal{O}_V$ induces an injective ring homomorphism from $\overline{k}[V]_{M_P}$ to $\mathcal{O}_P$. It is also surjective as every function in $\mathcal{O}_P$ is an element of $\overline{k}[V]_{M_P}$, and so $\overline{k}[V]_{M_P} \cong \mathcal{O}_P$. Now $\overline{k}[V]$ and $\overline{k}[V]_{M_P}$ have isomorphic quotient fields, which have the same transcendence degree over $\overline{k}$. By Theorem A.2.2(i) we obtain $\dim(\overline{k}[V]) = \dim(\overline{k}[V]_{M_P})$, thus $\dim(V) = \dim(\mathcal{O}_P)$ by Theorem 2.3.18.

(iv) By the above, the quotient field of $\overline{k}[V]$ is isomorphic to the quotient field of $\mathcal{O}_P$. It is easy to see that the quotient field of $\mathcal{O}_P$ is the function field $\overline{k}(V)$. Since $\overline{k}[V]$ is a finitely generated algebra over $\overline{k}$, the function field $\overline{k}(V)$ is also a finitely generated field over $\overline{k}$. Furthermore, by Theorems 2.3.18 and A.2.2(i), $\dim(V)$ is equal to the transcendence degree of $\overline{k}(V)$ over $\overline{k}$.

(i) Using again the injective map ω as well as (iii), we obtain

$$\overline{k}[V] \subseteq \mathcal{O}_V \subseteq \bigcap_{P \in V} \mathcal{O}_P = \bigcap_{P \in V} \overline{k}[V]_{M_P} = \overline{k}[V].$$

This proves (i).   □

Remark 2.4.14.

(i) Let V/k be an affine variety. The *k-rational function field* $k(V)$ of V/k is defined to be the quotient field of $k[V]$.

(ii) If $V \subseteq \mathbf{P}^n$ is a projective variety, then we can find θ_i in (2.2) such that $V \cap \theta_i^{-1}(\mathbf{A}^n) = V \cap U_i$ is not empty. Thus, by Lemma 2.4.10 the function field $\overline{k}(V)$ is the same as $\overline{k}(V \cap U_i) = \overline{k}(V \cap \mathbf{A}^n)$, if we identify U_i with $\mathbf{A}^n$.

(iii) If $V \subseteq \mathbf{P}^n$ is a projective variety defined over k, then $V \cap U_i \neq \varnothing$ for some i and $V \cap U_i = V \cap \mathbf{A}^n$ is an affine variety defined over k. Thus, we can define the *k-rational function field* $k(V)$ to be the k-rational function field $k(V \cap \mathbf{A}^n)$.

We have defined the Galois action of $\mathrm{Gal}(\overline{k}/k)$ on the polynomial ring $\overline{k}[X]$ in Section 2.2. This action can be defined on $\overline{k}[V]$ for an affine variety V/k because of the following result.

Lemma 2.4.15. Let V/k be an affine variety. Then for any polynomials $a, b \in \overline{k}[X]$ with $a - b \in I(V)$ and any automorphism $\sigma \in \mathrm{Gal}(\overline{k}/k)$, we have $\sigma(a)(P) = \sigma(b)(P)$ for all $P \in V$.

Proof. By Proposition 2.2.3, the conjugate $\sigma^{-1}(P)$ is also in V. Thus by Lemma 2.2.2,

$$\sigma(a)(P) = \sigma(a)(\sigma(\sigma^{-1}(P))) = \sigma(a(\sigma^{-1}(P)))$$
$$= \sigma(b(\sigma^{-1}(P))) = \sigma(b)(\sigma(\sigma^{-1}(P))) = \sigma(b)(P).$$

This completes the proof.   □

We can also extend the Galois action to the quotient field $\overline{k}(V)$ of $\overline{k}[V]$ by defining $\sigma(f/g) = \sigma(f)/\sigma(g)$ for $f, g \in \overline{k}[V]$ with $g \neq 0$.

We now want to look at the relationship between $k[V]$ (respectively $k(V)$) and $\overline{k}[V]$ (respectively $\overline{k}(V)$). For two polynomials $f, g \in k[X]$, we note that

$f - g \in I(V)$ if and only if $f - g \in I(V/k)$. Thus, $k[V]$ is naturally embedded into $\overline{k}[V]$, and therefore $k(V)$ is a subfield of $\overline{k}(V)$.

Theorem 2.4.16. Let V/k be an affine variety. Then

$$\overline{k}[V] = \overline{k} \cdot k[V], \qquad \overline{k}(V) = \overline{k} \cdot k(V).$$

Proof. Since $I(V)$ can be generated by polynomials in $k[X]$, it is easy to see that $I(V) = I(V/k)\overline{k}[X]$. Hence,

$$\overline{k}[V] = \overline{k}[X]/I(V) = \overline{k}[X]/(I(V/k)\overline{k}[X]) = \overline{k} \cdot k[V],$$

and the quotient field of $\overline{k} \cdot k[V]$ is $\overline{k} \cdot k(V)$.   □

Theorem 2.4.17. Let V/k be an affine variety. Then

$$k[V] = \{f \in \overline{k}[V] : \sigma(f) = f \quad \text{for all } \sigma \in \mathrm{Gal}(\overline{k}/k)\},$$

$$k(V) = \{h \in \overline{k}(V) : \sigma(h) = h \quad \text{for all } \sigma \in \mathrm{Gal}(\overline{k}/k)\}.$$

Proof. Denote by A the set $\{f \in \overline{k}[V] : \sigma(f) = f$ for all $\sigma \in \mathrm{Gal}(\overline{k}/k)\}$. It is clear that $k[V] \subseteq A$. Next, for an element f in A, by Theorem 2.4.16, f can be written as $\sum_{j=1}^{r} a_j f_j$ for some $a_j \in \overline{k}$ and $f_j \in k[V]$. We may assume that $\{f_1, \ldots, f_s\}$ is a $\overline{k}$-basis of the space $\langle f_1, \ldots, f_r \rangle$. Then $f = \sum_{j=1}^{s} b_j f_j$ for some $b_j \in \overline{k}$. Thus, for $\sigma \in \mathrm{Gal}(\overline{k}/k)$,

$$\sigma(f) = \sum_{j=1}^{s} \sigma(b_j) f_j = f = \sum_{j=1}^{s} b_j f_j,$$

that is, $\sum_{j=1}^{s}(\sigma(b_j) - b_j)f_j = 0$. This implies that $\sigma(b_j) = b_j$ for $1 \leq j \leq s$ and every $\sigma \in \mathrm{Gal}(\overline{k}/k)$. Hence, $b_j \in k$ for $1 \leq j \leq s$ and f is an element of $k[V]$. So $k[V] = A$.

The same arguments can be applied to prove the second part.   □

We recall from Section 1.5 that the full constant field of a function field over k consists of the elements of the function field that are algebraic over k.

Corollary 2.4.18. Let V/k be an affine variety of positive dimension. Then k is the full constant field of the function field $k(V)$.

Proof. If $a \in k(V)$ is algebraic over k, then $a \in \bar{k}$. By Theorem 2.4.17, $\sigma(a) = a$ for all $\sigma \in \mathrm{Gal}(\bar{k}/k)$. Hence, a is an element of k. $\qquad\square$

Remark 2.4.19. If V/k is a projective variety over k, then it is clear from the above that we have:

(i) $\bar{k}(V) = \bar{k} \cdot k(V)$;
(ii) $k(V) = \{h \in \bar{k}(V) : \sigma(h) = h \text{ for all } \sigma \in \mathrm{Gal}(\bar{k}/k)\}$;
(iii) k is the full constant field of the function field $k(V)$ if V has positive dimension.

2.5 Morphisms and Rational Maps

In the following, we consider maps between varieties, which, in a sense, take regular functions to regular functions. Recall that regular functions were introduced in Definition 2.4.2. In this section, we use the term variety to stand for quasi-affine variety or quasi-projective variety.

Definition 2.5.1. Let V and W be two varieties over $\bar{k}$. A *morphism* $\varphi : V \to W$ is a continuous map such that for every open subset U of W with $\varphi^{-1}(U) \neq \varnothing$ and every regular function f on U, the composite function $f \circ \varphi$ is a regular function on $\varphi^{-1}(U)$.

Proposition 2.5.2.

(i) Let $V_1 \subseteq \mathbf{A}^n$ and $V_2 \subseteq \mathbf{A}^m$ be two affine varieties. If $f_1, \ldots, f_m \in \bar{k}[X] = \bar{k}[x_1, \ldots, x_n]$ are such that

$$(f_1(P), \ldots, f_m(P)) \in V_2$$

for all points $P \in V_1$, then the map defined by

$$\varphi : V_1 \to V_2, \quad P \mapsto (f_1(P), \ldots, f_m(P)),$$

is a morphism.

(ii) Let $V_1 \subseteq \mathbf{P}^n$ and $V_2 \subseteq \mathbf{P}^m$ be two projective varieties. If $f_0, f_1, \dots,$ f_m are homogeneous polynomials of the same degree in $\bar{k}[X] = \bar{k}[x_0, x_1, \dots, x_n]$ such that

$$[f_0(P), f_1(P), \dots, f_m(P)] \in V_2$$

for all points $P \in V_1$, then the map defined by

$$\varphi : V_1 \to V_2, \qquad P \mapsto [f_0(P), f_1(P), \dots, f_m(P)],$$

is a morphism.

Proof. We prove only part (i). Assume that W is a closed subset of V_2. Then there exists an ideal $I \subseteq \bar{k}[y_1, \dots, y_m]$ such that $W = Z(I)$. Let J be the ideal of $\bar{k}[X]$ generated by

$$\{g(f_1(X), \dots, f_m(X)) : g \in I\}.$$

Then it is easy to verify that $\varphi^{-1}(W) = Z(J) \cap V_1$. Hence, $\varphi^{-1}(W)$ is a closed subset of V_1. This means that φ is continuous.

Now let U be an open subset of V_2 with $\varphi^{-1}(U) \neq \varnothing$ and let f be a regular function on U. For any point $P \in \varphi^{-1}(U)$, let U' be an open subset of U containing $\varphi(P)$ such that $f = a/b$ for some $a, b \in \bar{k}[y_1, \dots, y_m]$ and $b(Q) \neq 0$ for all $Q \in U'$. Then $\varphi^{-1}(U')$ is an open subset of $\varphi^{-1}(U)$ containing P and

$$f \circ \varphi = \frac{a \circ \varphi}{b \circ \varphi} = \frac{a(f_1(X), \dots, f_m(X))}{b(f_1(X), \dots, f_m(X))}.$$

It is clear that $(b \circ \varphi)(R) = b(\varphi(R)) \neq 0$ for all $R \in \varphi^{-1}(U')$ since $\varphi(R)$ belongs to U'. This shows that $f \circ \varphi$ is regular on $\varphi^{-1}(U)$.   □

Example 2.5.3.

(i) Assume that the characteristic of k is odd and let V be the affine variety over k defined by

$$x^2 + y^2 = 1.$$

Then the map

$$\varphi : V \to \mathbf{A}^1, \qquad (x, y) \mapsto (x + 1)y,$$

is a morphism.

(ii) Assume that the characteristic of $\mathbb{F}_q$ is odd and let V be the projective variety over $\mathbb{F}_q$ defined by

$$y^2z = x^3 + z^3.$$

Then the map

$$\varphi : V \to \mathbf{P}^2, \quad [x, y, z] \mapsto [x^2, xy, z^2],$$

is a morphism.

Proposition 2.5.4. Let V be a variety.

 (i) A regular function f on V is continuous when $\bar{k}$ is identified with $\mathbf{A}^1$.
(ii) Let $W \subseteq \mathbf{A}^n$ be an affine variety. A map $\psi : V \to W$ is a morphism if and only if $x_i \circ \psi$ is a regular function on V for each i, where x_i, $i = 1, \ldots, n$, are the coordinate functions on $\mathbf{A}^n$.

Proof.

 (i) It suffices to show that $f^{-1}(S)$ is a closed subset of V for every closed subset S of $\mathbf{A}^1$. We know from Example 2.2.7 that a closed subset of $\mathbf{A}^1$ is either $\mathbf{A}^1$ itself or a finite set. If $S = \mathbf{A}^1$, then $f^{-1}(S) = V$ is closed. Otherwise, S is a finite set and, hence, we may assume that $S = \{a\}$ for a single point a of $\mathbf{A}^1$. Let $Y \subseteq V$ be an open set on which f can be represented as g/h with $g, h \in \bar{k}[X]$ and h nowhere 0 on Y. So, $f^{-1}(a) \cap Y = \{P \in Y : g(P)/h(P) = a\} = Z(g - ah) \cap Y$ is a closed subset of Y. Since f is regular on V and V is compact, there is a finite open cover $\{Y_j\}$ of V such that for each Y_j we have $f^{-1}(a) \cap Y_j$ closed in Y_j. Thus, $f^{-1}(a)$ is closed.
(ii) If ψ is a morphism, then $x_i \circ \psi$ is a regular function on V by the definition of a morphism. Conversely, assume that $x_i \circ \psi$ is regular on V for all $1 \le i \le n$. Then $g \circ \psi$ is regular on V for any polynomial $g \in \bar{k}[X]$. Since by (i) regular functions are continuous and a closed subset of W is defined by the vanishing of polynomial functions, ψ^{-1} takes closed sets to closed sets, that is, ψ is continuous. Finally, since regular functions on nonempty open subsets of W are locally quotients of polynomials, $f \circ \psi$ is regular

on $\psi^{-1}(U)$ for any regular function f on any open subset U of W with $\psi^{-1}(U) \neq \varnothing$. This completes the proof.   □

Theorem 2.5.5. Let V be a variety and let W be an affine variety. Then there is a natural bijection

$$\varrho : \mathrm{Mor}(V, W) \to \mathrm{Hom}(\overline{k}[W], \mathcal{O}_V),$$

where Mor on the left-hand side means morphisms of varieties, while Hom on the right-hand side stands for homomorphisms of $\overline{k}$-algebras.

Proof. A morphism from V to W clearly carries a regular function on W to a regular function on V. Hence, it induces a map from $\mathcal{O}_W$ to $\mathcal{O}_V$, which is a homomorphism of $\overline{k}$-algebras. By Theorem 2.4.13(i), $\mathcal{O}_W$ is isomorphic to $\overline{k}[W]$. This defines a map ϱ.

Now suppose that we are given a homomorphism $\alpha : \overline{k}[W] \to \mathcal{O}_V$ of $\overline{k}$-algebras. We have $\overline{k}[W] = \overline{k}[x_1, \ldots, x_n]/I(W)$ by definition. Let $\bar{x}_i$ be the residue class of x_i in $\overline{k}[W]$ for $i = 1, \ldots, n$. With $f_i := \alpha(\bar{x}_i) \in \mathcal{O}_V$ we define the map $\delta : V \to \mathbf{A}^n$ by $\delta(P) = (f_1(P), \ldots, f_n(P))$ for $P \in V$. It is easy to verify that $\delta(V) \subseteq W$ and $\varrho(\delta) = \alpha$. By Proposition 2.5.4(ii), δ is a morphism from V to W.   □

Let V_1 and V_2 be two projective varieties (or affine varieties, respectively). If $\varphi : V_1 \to V_2$ is a morphism defined by polynomials as in Proposition 2.5.2, we simply denote φ by $[f_0(X), f_1(X), \ldots, f_m(X)]$ (or $(f_1(X), \ldots, f_m(X))$, respectively). We note also that two varieties V and W are called *isomorphic* if there are morphisms $\varphi : V \to W$ and $\psi : W \to V$ such that $\psi \circ \varphi$ and $\varphi \circ \psi$ are the identity maps on V and W, respectively.

Corollary 2.5.6. Let $V \subseteq \mathbf{A}^n$ and $W \subseteq \mathbf{A}^m$ be two affine varieties. Then:

  (i) morphisms from V to W are the same as polynomial maps;
 (ii) V and W are isomorphic if and only if $\overline{k}[V]$ and $\overline{k}[W]$ are isomorphic as $\overline{k}$-algebras.

Proof.

  (i) We note that a polynomial map gives a morphism by Proposition 2.5.2(i). Conversely, let φ be a morphism from V to W. By

Theorems 2.5.5 and 2.4.13(i), there exists a homomorphism of $\overline{k}$-algebras

$$\lambda : \overline{k}[W] = \overline{k}[y_1, \ldots, y_m]/I(W) \to \overline{k}[V] = \overline{k}[x_1, \ldots, x_n]/I(V)$$

induced by φ. Then it is easy to see that the polynomial map $(\lambda(\bar{y}_1), \ldots, \lambda(\bar{y}_m))$ is the morphism φ, where $\bar{y}_j$ is the residue class of y_j in $\overline{k}[W]$ for $j = 1, \ldots, m$.

(ii) This result is a direct consequence of Theorems 2.5.5 and 2.4.13(i).

$\square$

Proposition 2.5.7. Let $\varphi : V_1 \to V_2$ be a morphism of projective varieties. If there exist a nonempty open subset U of V_1 and homogeneous polynomials $f_0(X), f_1(X), \ldots, f_m(X)$ of the same degree such that

$$\varphi(Q) = [f_0(Q), f_1(Q), \ldots, f_m(Q)] \quad \text{for all } Q \in U,$$

then for any point $P \in V_1$ there exist homogeneous polynomials $h_0(X)$, $h_1(X), \ldots, h_m(X)$ of the same degree such that

$$\varphi(R) = [h_0(R), h_1(R), \ldots, h_m(R)]$$

for all points R in a suitable neighborhood of P, with $f_i h_j - f_j h_i \in I(V_1)$ for all $0 \le i, j \le m$.

Proof. Let V_1 and V_2 be varieties in $\mathbf{P}^n$ and $\mathbf{P}^m$, respectively. Let $\{U_i\}_{i=0}^{n}$ and $\{U_j'\}_{j=0}^{m}$ be the open covers of $\mathbf{P}^n$ and $\mathbf{P}^m$, respectively, in Example 2.3.4. We may assume that $P \in \varphi^{-1}(V_2 \cap U_0')$. Assume also without loss of generality that the first coordinate of P is not zero, that is, $P \in Y := \varphi^{-1}(V_2 \cap U_0') \cap U_0$. Then Y is a nonempty open subset of V_1 and can also be considered as an affine variety. Denote by f_i^* the polynomial $f_i(1, y_1, \ldots, y_n)$ and let $K = Z(f_0^*, \ldots, f_m^*)$. Then $J := (Y \setminus K) \cap U$ is a nonempty open subset of Y (see Theorem A.1.9(iv)) and we obviously have $\varphi|_J = [f_0, f_1, \ldots, f_m]$.

By Corollary 2.5.6(i), φ is a polynomial map when restricted to Y. Hence, we must have

$$\varphi|_Y = [1, g_1, \ldots, g_m]$$

for some polynomials g_i. Let e be the largest degree of the g_i and put

$$h_0(X) = x_0^e, \qquad h_i(X) = x_0^e g_i(x_1/x_0, \ldots, x_n/x_0) \quad \text{for } 1 \le i \le m.$$

Then it is clear that

$$\varphi|_Y = [h_0, h_1, \ldots, h_m].$$

Thus, when restricting to J, we get

$$\varphi|_J = [f_0, f_1, \ldots, f_m] = [h_0, h_1, \ldots, h_m].$$

Hence, we must have $(f_i h_j)(Q) = (f_j h_i)(Q)$ for all $0 \le i, j \le m$ and $Q \in J$, that is, $J \subseteq Z_h(f_i h_j - f_j h_i)$. Since $Z_h(f_i h_j - f_j h_i)$ is a closed subset of $\mathbf{P}^n$ and the closure of J in V_1 is V_1 (see Theorem A.1.9(i)), we obtain $V_1 \subseteq Z_h(f_i h_j - f_j h_i)$. Thus, $f_i h_j - f_j h_i \in I(V_1)$ since $f_i h_j - f_j h_i$ is a homogeneous polynomial. $\qquad\square$

Example 2.5.8. Consider the projective variety V over $\mathbb{F}_q$ with q odd defined by

$$x^2 + y^2 = z^2.$$

Then the map

$$\varphi : V \to \mathbf{P}^1, \qquad [x, y, z] \mapsto [x + z, y],$$

is regular on $V \setminus \{[1, 0, -1]\}$ by Example 2.5.3(i). Using the congruence

$$(x + z)(x - z) \equiv -y^2 \ (\text{mod } I(V)),$$

we have

$$[(x + z)(x - z), y(x - z)] = [-y^2, y(x - z)] = [-y, x - z].$$

Thus, with $\varphi([1, 0, -1]) = [0, 1]$, φ is a morphism on V.

Definition 2.5.9. Let V_i/k for $i = 1, 2$ be affine (or projective) varieties defined over k. A morphism $\varphi : V_1 \rightarrow V_2$ is *defined over k*, or a *k-morphism*, if $\sigma \circ \varphi = \varphi \circ \sigma$ for all $\sigma \in \mathrm{Gal}(\bar{k}/k)$.

Example 2.5.10.

(i) In the affine case, if all polynomials $f_i(X)$ belong to $k[X]$, then the morphism $(f_1(X), \ldots, f_m(X))$ is defined over k.
(ii) If $\varphi = [f_0, f_1, \ldots, f_m]$ is a morphism from the projective variety V_1/k to the projective variety V_2/k with $f_i \in k(V_1)$ for $i = 0, 1, \ldots, m$, then φ is a k-morphism.

Proposition 2.5.11. Let V be an affine (respectively projective) variety defined over k. Suppose that the characteristic of k is a prime p. Let $q > 1$ be a power of p and put

$$I^{(q)} := \left\{ f^{(q)} : \ f \in I(V) \right\},$$

where $f^{(q)}$ is the polynomial obtained from f by raising each coefficient of f to the qth power. Then $I^{(q)}$ is an ideal of $\bar{k}[X]$ and $V^{(q)} := Z(I^{(q)})$ (respectively $Z_h(I^{(q)})$) is an affine (respectively projective) variety defined over k. Furthermore, we have $\dim(V^{(q)}) = \dim(V)$.

Proof. We consider only the affine case. For any $f, g \in I(V)$, we have

$$f^{(q)} + g^{(q)} = (f + g)^{(q)} \in I^{(q)}.$$

As each element α of $\bar{k}$ is equal to β^q for some $\beta \in \bar{k}$, there exists, for any given $h \in \bar{k}[X]$, a polynomial $e \in \bar{k}[X]$ such that $h = e^{(q)}$. Thus,

$$h f^{(q)} = e^{(q)} f^{(q)} = (ef)^{(q)} \in I^{(q)}.$$

Hence, $I^{(q)}$ is an ideal of $\bar{k}[X]$.

Let S be a subset of $k[X]$ such that the ideal $I(V)$ of $\bar{k}[X]$ is generated by S. Then it is clear that $I^{(q)}$ is generated by $S^{(q)}$. Therefore, $V^{(q)}$ is an algebraic set defined over k.

Using Proposition 2.2.8, it is easily seen that $I(V^{(q)}) = I^{(q)}$. Now let $f_1, g_1 \in \bar{k}[X]$ be two polynomials such that $f_1 g_1$ belongs to $I^{(q)}$.

Let $f_2, g_2 \in \overline{k}[X]$ be the polynomials satisfying

$$f_1 = f_2^{(q)}, \qquad g_1 = g_2^{(q)}.$$

Then

$$(f_2 g_2)^{(q)} = f_2^{(q)} g_2^{(q)} = f_1 g_1 \in I^{(q)},$$

and so $f_2 g_2 \in I(V)$. This implies that $f_2 \in I(V)$ or $g_2 \in I(V)$ since $I(V)$ is a prime ideal of $\overline{k}[X]$ by Theorem 2.3.8. This means that $f_1 = f_2^{(q)} \in I^{(q)}$ or $g_1 = g_2^{(q)} \in I^{(q)}$. Hence, $I^{(q)}$ is a prime ideal of $\overline{k}[X]$, that is, $V^{(q)}$ is an affine variety defined over k.

Note that for every irreducible closed subset W of V, the set $W^{(q)}$ is an irreducible closed subset of $V^{(q)}$. Conversely, for every irreducible closed subset U of $V^{(q)}$, there exists an irreducible closed subset W of V such that $W^{(q)} = U$. Using this fact, it is easy to show that $\dim(V) = \dim(V^{(q)})$. $\qquad\square$

Definition 2.5.12. Let V be a projective (or affine) variety over k and $q > 1$ a power of the characteristic of k. Then the morphism $\pi_q : V \to V^{(q)}$ defined by

$$[x_0, x_1, \ldots, x_n] \mapsto [x_0^q, x_1^q, \ldots, x_n^q]$$

(or $(x_1, \ldots, x_n) \mapsto (x_1^q, \ldots, x_n^q)$) is called the *qth-power Frobenius morphism.*

Theorem 2.5.13. Let $V/\mathbb{F}_q$ be a projective (or affine) variety. Then $V = V^{(q)}$ and the qth-power Frobenius morphism π_q is bijective and bicontinuous, that is, π_q is a homeomorphism of topological spaces. Furthermore, a point P of V is $\mathbb{F}_q$-rational if and only if $\pi_q(P) = P$.

Proof. For any point $P = [a_0, a_1, \ldots, a_n]$ in a closed point P of V, the point $P^{(q)} := [a_0^q, a_1^q, \ldots, a_n^q] = \pi_q(P)$ is still a point of P. On the other hand, it is clear that $P^{(q)}$ is a point of $V^{(q)}$. Thus, P is also a closed point of $V^{(q)}$. This means that $V \subseteq V^{(q)}$. Conversely, for any point $Q = [b_0, b_1, \ldots, b_n]$ of a closed point Q of $V^{(q)}$, there exist elements $c_i \in \overline{\mathbb{F}}_q$ such that $c_i^q = b_i$ for $0 \le i \le n$. Thus, the point $R := [c_0, c_1, \ldots, c_n]$ belongs to V. It is clear that both R and Q are in the closed point Q. Hence, Q is also a closed point of V.

By Definition 2.5.9, π_q is an $\mathbb{F}_q$-morphism from V to itself. From the above, we can easily see that π_q is bijective. It is straightforward to verify that π_q^{-1} is continuous.

The last statement is clear. $\qquad\square$

Proposition 2.5.14.

(i) Let $V_f = Z(f)$ be the hypersurface in $\mathbf{A}^n$ defined by an irreducible polynomial $f(X)$ over $\bar{k}$ with $X = (x_1, \ldots, x_n)$. Then the open set $\mathbf{A}^n \setminus V_f$ is isomorphic to the hypersurface $Z(x_{n+1} f(X) - 1)$ in $\mathbf{A}^{n+1}$. In particular, the open set $\mathbf{A}^n \setminus V_f$ is isomorphic to an affine variety.

(ii) A variety can be covered by a family of open affine sets. In other words, for any point P in the variety and any neighborhood U of P, there is an open affine set W such that $P \in W \subseteq U$.

Proof.

(i) Consider the map φ from $G := Z(x_{n+1} f(X) - 1)$ to $\mathbf{A}^n \setminus V_f$ defined by $(a_1, \ldots, a_{n+1}) \mapsto (a_1, \ldots, a_n)$. It is easy to verify that φ is a bijection. It is clear that φ is a morphism since it is a polynomial map. Since $\varphi^{-1} = (x_1, \ldots, x_n, 1/f(X))$ and $f(P) \neq 0$ for every $P \in G$, the map φ^{-1} is a morphism as well by Proposition 2.5.4(ii). Hence φ is an isomorphism.

(ii) It is clear that any variety V can be covered by a family of quasi-affine varieties (compare with Example 2.3.4 for the projective case). Hence, we may assume that V is a quasi-affine variety, that is, V is an open subset of an affine variety $Y \subseteq \mathbf{A}^n$. We may also assume that $U = V$. Let $C = Y \setminus V$; then C is a closed set in $\mathbf{A}^n$. If P is a point in V, then $P \notin C$ and thus by Proposition 2.2.8(ii) there is a polynomial $f \in I(C)$ such that $f(P) \neq 0$. Let H be the hypersurface $H = Z(f)$. Then $P \notin H$, hence, $P \in V \setminus (V \cap H)$, which is an open subset of V. Furthermore, $V \setminus (V \cap H)$ is a closed subset of $\mathbf{A}^n \setminus H$, which is an affine variety by (i). Hence, $V \setminus (V \cap H)$ is the required affine neighborhood of P. $\qquad\square$

Let V and W be two varieties over $\bar{k}$. Consider the set of pairs

$$\{\langle U, \varphi_U \rangle : \varnothing \neq U \subseteq V \text{ is open and } \varphi_U : U \to W \text{ is a morphism}\}.$$

Define a relation on the above set by

$$\langle U, \varphi_U \rangle \sim \langle S, \varphi_S \rangle \quad \text{if and only if } \varphi_U \text{ and } \varphi_S \text{ agree on } U \cap S. \quad (2.3)$$

Note that $U \cap S$ is a nonempty open subset of V by Theorem A.1.9(iv).

Lemma 2.5.15.

(i) Let V and W be two varieties and let φ and ψ be two morphisms from V to W. If there exists a nonempty open subset U of V such that φ and ψ agree on U, then $\varphi = \psi$.

(ii) The relation $\sim$ defined in (2.3) is an equivalence relation.

Proof.

(i) We have to show that $\varphi(P) = \psi(P)$ for every $P \in V$. Since this is a local property and a variety can always be covered by open subsets of affine varieties by Proposition 2.5.14(ii), we may assume that both V and W are affine varieties. Hence, by Corollary 2.5.6(i), $\varphi = (f_1(X), \ldots, f_m(X))$ and $\psi = (g_1(X), \ldots, g_m(X))$ for some polynomials $f_i, g_j \in \overline{k}[X]$. As $\varphi(Q) = \psi(Q)$ for all points $Q \in U$, we have $U \subseteq Z(f_1 - g_1, \ldots, f_m - g_m)$. So $Z(f_1 - g_1, \ldots, f_m - g_m)$ contains the closure of U, which is V by Theorem A.1.9(i). Therefore $\varphi(P) = \psi(P)$ for any $P \in V$.

(ii) Let $\langle U, \varphi_U \rangle \sim \langle S, \varphi_S \rangle$ and $\langle S, \varphi_S \rangle \sim \langle T, \varphi_T \rangle$. Then φ_U and φ_S agree on $U \cap S$ and φ_S and φ_T agree on $S \cap T$. Thus, φ_U and φ_T agree on $U \cap S \cap T$. By part (i), φ_U and φ_T agree on $U \cap T$, that is, $\langle U, \varphi_U \rangle \sim \langle T, \varphi_T \rangle$. The remaining properties of an equivalence relation are trivial. □

On the basis of Lemma 2.5.15(ii), it is meaningful to introduce the following definition.

Definition 2.5.16. Let V and W be two varieties over $\overline{k}$. A *rational map* φ from V to W is an equivalence class of pairs $\langle U, \varphi_U \rangle$ defined in (2.3).

Remark 2.5.17. A rational map φ from V to W is not in general a map as some elements of V may not have images in W.

Example 2.5.18. The rational map

$$\varphi : \mathbf{P}^2 \to \mathbf{P}^2, \qquad \varphi = [x^2, xy, z^2],$$

is a morphism from $\mathbf{P}^2 \setminus \{[0, 1, 0]\}$ to $\mathbf{P}^2$. It is not a map as $\varphi([0, 1, 0])$ is not well defined.

Definition 2.5.19. The *domain*, denoted by $\mathrm{dom}(\varphi)$, of a rational map φ from V to W is the union of all open subsets U of V such that some $\langle U, \varphi_U \rangle$ belongs to the equivalence class of this rational map.

Example 2.5.20. The domain of the rational map in Example 2.5.18 is $\mathbf{P}^2 \setminus \{[0, 1, 0]\}$.

Remark 2.5.21. It is clear that for a rational map φ from V to W, the domain $\mathrm{dom}(\varphi)$ is a nonempty open subset of V, hence, $\mathrm{dom}(\varphi)$ is a (quasi-affine or quasi-projective) variety and φ is a morphism from $\mathrm{dom}(\varphi)$ to W.

Definition 2.5.22. A rational map φ from V to W is *dominant* if $\varphi(\mathrm{dom}(\varphi))$ is dense in W.

For instance, the rational map in Example 2.5.18 is dominant. It is easy to see that a rational map φ from V to W is dominant if and only if for every pair $\langle U, \varphi_U \rangle$ in this class, $\varphi_U(U)$ is dense in W.

Definition 2.5.23. A rational map φ from V to W is said to be *birational* if there exist a nonempty open subset U of V and a nonempty open subset S of W such that φ is an isomorphism from U to S. In this case, V and W are said to be *birationally equivalent*.

It is clear that if φ is a birational map from V to W, then it is dominant. The following theorem and its corollaries are major results of this section. We refer to Definition 2.4.8 for the definition of the function field $\bar{k}(V)$ of a variety V.

Theorem 2.5.24. Let V and W be two varieties over $\bar{k}$. Then there is a one-to-one correspondence between dominant rational maps from V to W and $\bar{k}$-algebra homomorphisms from $\bar{k}(W)$ to $\bar{k}(V)$.

Proof. Let $\psi = \langle U, \psi_U \rangle$ be a dominant rational map from V to W. Let $\overline{(Y, f)} \in \overline{k}(W)$, where Y is a nonempty open subset of W and f is regular on Y. Since $\psi_U(U)$ is dense in W and ψ_U is a morphism, we conclude that $\psi_U^{-1}(Y)$ is a nonempty open subset of V. Again by the definition of a morphism, the composition $f \circ \psi_U$ is a regular function on $\psi_U^{-1}(Y)$, that is, we get a map

$$\kappa : \overline{k}(W) \to \overline{k}(V), \qquad \overline{(Y, f)} \mapsto \overline{(\psi_U^{-1}(Y), f \circ \psi_U)}.$$

It is easy to verify that κ is a $\overline{k}$-algebra homomorphism from $\overline{k}(W)$ to $\overline{k}(V)$.

Conversely, assume that we have a $\overline{k}$-algebra homomorphism κ from $\overline{k}(W)$ to $\overline{k}(V)$. Since W can be covered by open affine sets according to Proposition 2.5.14(ii), we may assume that W is affine. Let $y_1, \ldots, y_m$ be the coordinate functions of $\overline{k}[W]$. Then for each $i = 1, \ldots, m$, we get an element $\kappa(y_i) = \overline{(U_i, f_i)}$ of $\overline{k}(V)$. If U is the intersection of all U_i, then U is a nonempty open subset of V and f_i is regular on U for each i. Thus, we obtain an injective homomorphism from $\overline{k}[W]$ to $\mathcal{O}_U$. By Theorem 2.5.5, this induces a morphism ψ from U to W, which is a dominant rational map from V to W. $\qquad\square$

Corollary 2.5.25. Two varieties over $\overline{k}$ are birationally equivalent if and only if their function fields are isomorphic as $\overline{k}$-algebras.

Corollary 2.5.26. Every curve is birationally equivalent to a plane curve.

Proof. The function field $\overline{k}(\mathcal{X})$ of a curve $\mathcal{X}$ has transcendence degree 1 over $\overline{k}$ by Theorem 2.4.13(iv) and Remark 2.4.14(ii), and so $\overline{k}(\mathcal{X}) = \overline{k}(y_1, y_2)$ for some y_1, y_2, where y_1 is transcendental over $\overline{k}$ and y_2 is algebraic over $\overline{k}(y_1)$ (compare with [51, p. 27] and [117, Proposition III.9.2]). Consider the natural homomorphism from the polynomial ring $\overline{k}[x_1, x_2]$ to $\overline{k}[y_1, y_2]$. Then the kernel I must be a prime ideal. Thus, $\mathcal{Y} = Z(I)$ is a plane curve. By Corollary 2.5.25, $\mathcal{Y}$ is birationally equivalent to $\mathcal{X}$. $\qquad\square$

3   Algebraic Curves

Algebraic curves, that is, algebraic varieties of dimension 1, are crucial for the applications that will be discussed in Chapters 5 and 6. In the present chapter, we emphasize the interplay between algebraic curves and function fields, which has become a powerful tool in both the theory and the applications.

The important family of nonsingular or smooth curves is introduced in Section 3.1. For nonsingular projective curves we establish a link between points of the curve and valuations of the corresponding function field. This link is strengthened in Section 3.2 where it is shown that nonsingular projective curves and algebraic function fields of one variable are basically equivalent mathematical objects. The theory of divisors is developed in Section 3.3 in the language of function fields. Sections 3.4 and 3.5 set the stage for the Riemann-Roch theorem by introducing Riemann-Roch spaces and the fundamental concept of genus. Section 3.6 achieves the proof of the Riemann-Roch theorem via an approach based on adèles and Weil differentials. The special family of elliptic curves is treated in Section 3.7. The intriguing relationships between nonsingular projective curves over finite fields and global function fields are summarized in Section 3.8.

The books of Bump [13] and Hartshorne [51] on algebraic geometry contain also a good amount of material on algebraic curves. Fulton [35] is a classical monograph devoted specifically to algebraic curves. The recent book of Hirschfeld, Korchmáros, and Torres [55] emphasizes algebraic curves over finite fields.

We continue to assume that k is a perfect field, and on several occasions we highlight the special case where k is a finite field, which will be important in the following chapters.

3.1 Nonsingular Curves

We have already defined (algebraic) curves briefly in Section 2.3. We record the definition again for the convenience of the reader.

Definition 3.1.1. Let k be a perfect field. An affine (respectively projective) variety of dimension 1 defined over k is called an *affine* (respectively *projective*) *curve* over k.

We speak of a curve (with no adjective) if it can be either affine or projective. If we want to emphasize the field k, then we write $\mathcal{X}/k$ for a curve $\mathcal{X}$ (defined) over k.

Example 3.1.2.

(i) The projective variety defined over k with char$(k) = 2$ given by

$$zy^2 + yz^2 = x^3$$

is a projective curve.

(ii) The affine variety defined over k with char$(k) \neq 2$ given by

$$y^2 = f(x)$$

is an affine curve, where $f \in k[x]$ is a squarefree polynomial of positive degree.

Definition 3.1.3. Let $\mathcal{X} \subseteq \mathbf{A}^n$ be an affine curve and let $f_1, \ldots, f_m \in \bar{k}[X] = \bar{k}[x_1, \ldots, x_n]$ be a set of generators of the ideal $I(\mathcal{X})$. Then $\mathcal{X}$ is said to be *nonsingular* (or *smooth*) at a point P of $\mathcal{X}$ if the $m \times n$ *Jacobian matrix*

$$\left(\frac{\partial f_i}{\partial x_j}(P) \right)_{1 \leq i \leq m, \, 1 \leq j \leq n}$$

at P has rank $n - 1$. Otherwise, $\mathcal{X}$ is said to be *singular* at P. If $\mathcal{X}$ is smooth at every point of $\mathcal{X}$, then we say that $\mathcal{X}$ is a *nonsingular* (or *smooth*) *affine curve*.

Alternatively, if $\mathcal{X}$ is nonsingular (or smooth) at P, then we speak also of a nonsingular (or smooth) point P of $\mathcal{X}$. Similarly, we may speak of singular points of $\mathcal{X}$.

Example 3.1.4. Let $\mathcal{X}$ be an affine plane curve defined by

$$f(x, y) = 0$$

for a polynomial $f \in k[x, y]$. By Definition 3.1.3, a point P of $\mathcal{X}$ is smooth if and only if

$$\left(\frac{\partial f}{\partial x}(P), \frac{\partial f}{\partial y}(P) \right) \neq (0, 0).$$

In other words, all singular points (a, b) of $\mathcal{X}$ are solutions of the system of equations

$$\begin{cases} f(a, b) = 0, \\ \frac{\partial f}{\partial x} \big|_{(a,b)} = 0, \\ \frac{\partial f}{\partial y} \big|_{(a,b)} = 0. \end{cases}$$

If $P = (a, b)$ is a smooth point of $\mathcal{X}$, then the line

$$\frac{\partial f}{\partial x}(P)(x - a) + \frac{\partial f}{\partial y}(P)(y - b) = 0$$

is called the *tangent line* of $\mathcal{X}$ at P.

Theorem 3.1.5. Let P be a point of an affine curve $\mathcal{X}$. Then $\mathcal{X}$ is smooth at P if and only if the local ring $\mathcal{O}_P$ at P is a discrete valuation ring in the sense of Definition A.3.1(ii).

Proof. Let $\mathcal{X} \subseteq \mathbf{A}^n$ and $P = (a_1, \ldots, a_n) \in \mathcal{X}$. Let A_P be the maximal ideal $(x_1 - a_1, \ldots, x_n - a_n)$ of $\overline{k}[X] = \overline{k}[x_1, \ldots, x_n]$. We define a $\overline{k}$-linear map $\theta : \overline{k}[X] \to \overline{k}^n$ by

$$f \mapsto \left(\frac{\partial f}{\partial x_1}(P), \ldots, \frac{\partial f}{\partial x_n}(P) \right).$$

It is clear that $\theta(x_i - a_i)$ for $i = 1, \ldots, n$ form a basis of $\overline{k}^n$ and that $\theta(A_P^2) = \{(0, \ldots, 0)\}$. Thus, θ induces an isomorphism $\theta' : A_P / A_P^2 \to \overline{k}^n$.

Now let $B = I(\mathcal{X})$ and let $f_1, \ldots, f_m$ be a set of generators of B. Then the rank of the Jacobian matrix $J := ((\partial f_i / \partial x_j)(P))$ is just the dimension of $\theta(B)$ as a subspace of $\overline{k}^n$. Using the isomorphism θ', this dimension is the same as the dimension of the subspace $(B + A_P^2)/A_P^2$ of A_P / A_P^2. On the other hand, the local ring $\mathcal{O}_P$ is obtained from $\overline{k}[X]$ by the procedure described in Theorem 2.4.13. Thus, if $\mathcal{M}_P$ is the maximal ideal of $\mathcal{O}_P$, then we have

$$\mathcal{M}_P / \mathcal{M}_P^2 \simeq A_P / (B + A_P^2).$$

Counting dimensions of vector spaces, we obtain $\dim_{\overline{k}}(\mathcal{M}_P/\mathcal{M}_P^2) +$ rank$(J) = n$. Note that $\mathcal{O}_P$ is Noetherian since $\overline{k}[X]$ is Noetherian by the Hilbert basis theorem and factor rings and localizations of Noetherian rings are Noetherian (see [29, Sections 1.4 and 2.1]). The desired result follows therefore from Definition 3.1.3 and Proposition A.3.4. □

Remark 3.1.6. From the proof of Theorem 3.1.5 we see that the Jacobian matrix has rank at most $n - 1$. Thus, the idea of Definition 3.1.3 is to call $\mathcal{X}$ nonsingular at P if the Jacobian matrix at P has the largest possible rank.

Theorem 3.1.5 provides an intrinsic characterization of smooth points. This characterization can be used to extend Definition 3.1.3 from affine to projective curves. Thus, we say that a projective curve $\mathcal{X}$ is *nonsingular* (or *smooth*) at $P \in \mathcal{X}$ if $\mathcal{O}_P$ is a discrete valuation ring. Otherwise, $\mathcal{X}$ is said to be *singular* at P. Similarly, we speak of a *nonsingular* (or *smooth*) *projective curve* $\mathcal{X}$ if $\mathcal{X}$ is smooth at every point of $\mathcal{X}$. The terminology mentioned in the paragraph following Definition 3.1.3 will be used in the same way for projective curves.

Theorem 3.1.7. The set of singular points of a curve $\mathcal{X}/k$ is finite.

Proof. By considering an open cover $\{\mathcal{Y}_i\}$ of $\mathcal{X}$ with each $\mathcal{Y}_i$ being an affine curve, we may assume that $\mathcal{X}$ is an affine curve.

By Remark 3.1.6, we know that the rank of the Jacobian matrix is always at most $n - 1$. Hence, the set Z of singular points of $\mathcal{X}$ is the set of points where the rank is less than $n - 1$. Thus, Z is the algebraic set defined by the ideal generated by $I(\mathcal{X})$ together with all determinants of $(n-1) \times (n-1)$ submatrices of the Jacobian matrix. Hence, Z is closed.

By Corollary 2.5.26, $\mathcal{X}$ is birationally equivalent to an affine plane curve. Thus, we may assume that $\mathcal{X}$ is an affine plane curve defined by a single polynomial equation $f(x, y) = 0$ with $f \in k[x, y]$ absolutely irreducible. By Example 3.1.4 and by distinguishing between the cases of characteristic 0 and positive characteristic of k, it is easily seen that Z is a proper closed subset of $\mathcal{X}$, and this leads to the desired result. □

Example 3.1.8. Consider the affine curve in Example 3.1.2(ii). Then it is easy to verify that this curve has no singular points, and so it is a nonsingular affine curve. Similarly, the curve in Example 3.1.2(i) is a nonsingular projective curve.

It is an important consequence of Remark 2.4.14(ii) and of Theorem 2.4.13 and its proof that the function field $\overline{k}(\mathcal{X})$ of a curve $\mathcal{X}/k$ is an algebraic function field of one variable over $\overline{k}$. A similar statement holds for the k-rational function field $k(\mathcal{X})$ of $\mathcal{X}$. By Corollary 2.4.18 and Remark 2.4.19, the full constant field of $k(\mathcal{X})$ is k.

Definition 3.1.9. For a nonsingular point P of a curve $\mathcal{X}$, a local parameter of the discrete valuation ring $\mathcal{O}_P(\mathcal{X})$ is called a *local parameter* or *uniformizing parameter* at P. Furthermore, the ord function of $\mathcal{O}_P(\mathcal{X})$ defined according to Definition A.3.5 is denoted by v_P.

Remark 3.1.10.

(i) As in Definition A.3.5, the function v_P can be extended to the quotient field $\overline{k}(\mathcal{X})$ of $\mathcal{O}_P(\mathcal{X})$. By what has been noted in Definition A.3.5, v_P satisfies the properties (1), (2), (3), and (4) in Definition 1.5.3. Since the nonzero elements of $\overline{k}$ are units of $\mathcal{O}_P(\mathcal{X})$, it follows that v_P satifies also the property (5) in Definition 1.5.3. Thus, v_P is a valuation of the function field $\overline{k}(\mathcal{X})/\overline{k}$.

(ii) If P and Q are two distinct nonsingular points of the projective curve $\mathcal{X}$, then the valuations v_P are v_Q are not equivalent. To see this, we may assume that $P = [a_0, a_1, \ldots, a_{n-1}, 1]$ and $Q = [b_0, b_1, \ldots, b_{n-1}, 1]$ with $a_0 \neq b_0$. Then $1/(x_0/x_n - a_0) \notin \mathcal{O}_P(\mathcal{X})$ and $1/(x_0/x_n - a_0) \in \mathcal{O}_Q(\mathcal{X})$. Hence, $\mathcal{O}_P(\mathcal{X}) \neq \mathcal{O}_Q(\mathcal{X})$, which implies that v_P and v_Q are not equivalent.

For two local rings A and B with $A \subseteq B$, we say that B *dominates* A if the maximal ideal of B contains the maximal ideal of A.

Lemma 3.1.11. Let $\mathcal{X}/k$ be a nonsingular projective curve and suppose that R is a discrete valuation ring with quotient field $\overline{k}(\mathcal{X})$. Then there exists a unique point P of $\mathcal{X}$ such that R dominates $\mathcal{O}_P(\mathcal{X})$.

Proof. Let ν be the ord map of R according to Definition A.3.5. Suppose that P and Q are two distinct points of $\mathcal{X}$ such that R dominates both $\mathcal{O}_P(\mathcal{X})$ and $\mathcal{O}_Q(\mathcal{X})$. By Remark 3.1.10(ii) and Theorem 1.5.18, there exists an element $z \in \overline{k}(\mathcal{X})$ such that $\nu_P(z) = 1$ and $\nu_Q(z) = -1$. Using Lemma A.3.2(ii), we deduce that $\nu(z) \geq 1$ and also $\nu(1/z) \geq 1$. This contradiction shows the uniqueness of the point P.

To show the existence of such a point P, we use the sets H_i and U_i in Example 2.3.4 and consider the map from $\mathbf{P}^{n-1}$ to the hyperplane H_n of $\mathbf{P}^n$ defined by $[x_0, \ldots, x_{n-1}] \mapsto [x_0, \ldots, x_{n-1}, 0]$. Thus, if $\mathcal{X}$ is contained in H_n, then $\mathcal{X}$ is isomorphic to a variety in $\mathbf{P}^{n-1}$. Hence we can assume that $\mathcal{X}$ is a projective curve in $\mathbf{P}^n$ such that $\mathcal{X}$ is not contained in any hyperplane H_i, that is, x_i/x_j are well-defined nonzero elements in the field $\overline{k}(\mathcal{X})$ for any $0 \leq i, j \leq n$. Put

$$m = \max_{0 \leq i, j \leq n} \nu\left(\frac{x_i}{x_j}\right).$$

Without loss of generality, we may assume that $m = \nu(x_0/x_n)$. So, $\nu(x_i/x_n) = m - \nu(x_0/x_i) \geq 0$. This implies that the coordinate ring $\overline{k}[\mathcal{Y}]$ of $\mathcal{Y} := \mathcal{X} \cap U_n$ is contained in R. Let M be the maximal ideal of R and let $I = M \cap \overline{k}[\mathcal{Y}]$. Then I is a prime ideal of $\overline{k}[\mathcal{Y}]$. We have $I \neq \{0\}$, for otherwise every nonzero element of $\overline{k}[\mathcal{Y}]$ is a unit of R, that is, $\overline{k}(\mathcal{X})$ is a subset of R since $\overline{k}(\mathcal{X})$ is the quotient field of $\overline{k}[\mathcal{Y}]$ (see Remark 2.4.14(ii) and Theorem 2.4.13(iv)). But then $R = \overline{k}(\mathcal{X})$, which is a contradiction since a discrete valuation ring cannot be a field because a field does not contain irreducible elements (compare with Definition A.3.1(ii)). Hence, I corresponds to a variety V properly contained in $\mathcal{Y}$. Let P be a point of V. Then $\{P\} \subseteq V \subsetneq \mathcal{Y}$, and since $\dim(\mathcal{Y}) = 1$, we must have $V = \{P\}$ by the definition of the dimension of a variety. It is easy to check that R dominates the local ring $\mathcal{O}_P(\mathcal{Y}) = \mathcal{O}_P(\mathcal{X})$. $\qquad\square$

Theorem 3.1.12. Let $\mathcal{X}/k$ be a nonsingular projective curve. Then the map

$$\rho: P \mapsto \mathcal{O}_P = \mathcal{O}_P(\mathcal{X})$$

yields a one-to-one correspondence between the points of $\mathcal{X}$ and the discrete valuation rings with quotient field $\bar{k}(\mathcal{X})$.

Proof. In Remark 3.1.10(ii), we have shown that $\mathcal{O}_P(\mathcal{X}) \neq \mathcal{O}_Q(\mathcal{X})$ for two distinct points P and Q of $\mathcal{X}$. Hence, ρ is injective.

Let R be a discrete valuation ring with quotient field $\bar{k}(\mathcal{X})$. By Lemma 3.1.11, there exists a point P of $\mathcal{X}$ such that $\mathcal{O}_P(\mathcal{X}) \subseteq R$. It follows from Proposition A.3.3 that $\mathcal{O}_P(\mathcal{X}) = R$. This means that ρ is surjective. □

The following lemma is shown in a straightforward manner. It refers to the Galois action on function fields introduced after Lemma 2.4.15.

Lemma 3.1.13. Let P and Q be two points of a projective curve $\mathcal{X}$ over k. Suppose that P and Q belong to the same k-closed point of $\mathcal{X}$ and that $\sigma \in \mathrm{Gal}(\bar{k}/k)$ satisfies $Q = \sigma(P)$. Then

$$\mathcal{O}_Q(\mathcal{X}) = \sigma(\mathcal{O}_P(\mathcal{X})).$$

Proposition 3.1.14. Let $\mathcal{X}/\mathbb{F}_q$ be a nonsingular projective curve. Then two points P and Q of $\mathcal{X}$ belong to the same $\mathbb{F}_q$-closed point if and only if

$$\mathcal{O}_P(\mathcal{X}) \cap \mathbb{F}_q(\mathcal{X}) = \mathcal{O}_Q(\mathcal{X}) \cap \mathbb{F}_q(\mathcal{X}).$$

Proof. Suppose P and Q belong to the same $\mathbb{F}_q$-closed point. By definition, there exists an element $\sigma \in \mathrm{Gal}(\overline{\mathbb{F}_q}/\mathbb{F}_q)$ such that $Q = \sigma(P)$. Thus, it follows from Lemma 3.1.13 that we have

$$\begin{aligned}
\mathcal{O}_Q(\mathcal{X}) \cap \mathbb{F}_q(\mathcal{X}) &= \sigma(\mathcal{O}_P(\mathcal{X})) \cap \mathbb{F}_q(\mathcal{X}) \\
&= \sigma(\mathcal{O}_P(\mathcal{X})) \cap \sigma(\mathbb{F}_q(\mathcal{X})) \\
&= \sigma(\mathcal{O}_P(\mathcal{X}) \cap \mathbb{F}_q(\mathcal{X})) \\
&= \mathcal{O}_P(\mathcal{X}) \cap \mathbb{F}_q(\mathcal{X}).
\end{aligned}$$

Conversely, assume that $\mathcal{O}_P(\mathcal{X}) \cap \mathbb{F}_q(\mathcal{X}) = \mathcal{O}_Q(\mathcal{X}) \cap \mathbb{F}_q(\mathcal{X})$. Suppose that P and Q do not belong to the same $\mathbb{F}_q$-closed point. Let P and Q be the two $\mathbb{F}_q$-closed points containing P and Q, respectively. By Theorem 1.5.18, we can find an element $z \in \overline{\mathbb{F}}_q(\mathcal{X})$ such that $v_{P'}(z) = -1$ for all $P' \in \mathsf{P}$ and $v_{Q'}(z) = 1$ for all $Q' \in \mathsf{Q}$. Let $\mathbb{F}_{q^m}$ be an extension field of $\mathbb{F}_q$ such that $\mathbb{F}_{q^m}$ contains the definition fields of P and Q (see Remark 2.1.10(iii)) and z belongs to $\mathbb{F}_{q^m}(\mathcal{X})$. Then the element $y := \prod_{\sigma \in \text{Gal}(\mathbb{F}_{q^m}/\mathbb{F}_q)} \sigma(z)$ is an element of $\mathbb{F}_q(\mathcal{X})$. Furthermore, we have

$$v_P(y) = \sum_{\sigma \in \text{Gal}(\mathbb{F}_{q^m}/\mathbb{F}_q)} v_P(\sigma(z)) = \sum_{\sigma \in \text{Gal}(\mathbb{F}_{q^m}/\mathbb{F}_q)} v_{\sigma(P)}(z) = -m.$$

This means that $y \notin \mathcal{O}_P(\mathcal{X}) \cap \mathbb{F}_q(\mathcal{X})$. On the other hand,

$$v_Q(y) = \sum_{\sigma \in \text{Gal}(\mathbb{F}_{q^m}/\mathbb{F}_q)} v_Q(\sigma(z)) = \sum_{\sigma \in \text{Gal}(\mathbb{F}_{q^m}/\mathbb{F}_q)} v_{\sigma(Q)}(z) = m,$$

that is, $y \in \mathcal{O}_Q(\mathcal{X}) \cap \mathbb{F}_q(\mathcal{X}) = \mathcal{O}_P(\mathcal{X}) \cap \mathbb{F}_q(\mathcal{X})$. This contradiction shows that P and Q are in the same $\mathbb{F}_q$-closed point.   □

The following theorem is one of the central results of this chapter. We refer to Section 1.5 for the concept of place of a function field.

Theorem 3.1.15. Let $\mathcal{X}/\mathbb{F}_q$ be a nonsingular projective curve. Then there exists a natural one-to-one correspondence between $\mathbb{F}_q$-closed points of $\mathcal{X}$ and places of $\mathbb{F}_q(\mathcal{X})$. Moreover, the degree of an $\mathbb{F}_q$-closed point is equal to the degree of the corresponding place.

Proof. For an $\mathbb{F}_q$-closed point P of $\mathcal{X}$, let P be a point in P and v_P the valuation of $\overline{\mathbb{F}_q}(\mathcal{X})$ introduced in Remark 3.1.10(i). Let $\mathcal{P}$ be the place of $\mathbb{F}_q(\mathcal{X})$ containing the restriction of v_P to $\mathbb{F}_q(\mathcal{X})$. The valuation ring of $\mathcal{P}$ is $\mathcal{O}_P(\mathcal{X}) \cap \mathbb{F}_q(\mathcal{X})$. If P' is another point in P, then define $\mathcal{P}'$ in an analogous way. Then using Proposition 3.1.14, we see that the valuation rings of $\mathcal{P}$ and $\mathcal{P}'$ are identical. It is now an immediate consequence of Theorem 1.5.18 that $\mathcal{P} = \mathcal{P}'$. Therefore the map ϕ sending each $\mathbb{F}_q$-closed point P of $\mathcal{X}$ to the corresponding place $\mathcal{P}$ of $\mathbb{F}_q(\mathcal{X})$ is well defined.

Let P and Q be two $\mathbb{F}_q$-closed points of $\mathcal{X}$ with $\phi(P) = \phi(Q)$. Choose $P \in$ P and $Q \in$ Q. Since the places $\phi(P)$ and $\phi(Q)$ are identical, they have the same valuation ring, that is, $\mathcal{O}_P(\mathcal{X}) \cap \mathbb{F}_q(\mathcal{X}) = \mathcal{O}_Q(\mathcal{X}) \cap \mathbb{F}_q(\mathcal{X})$. But then Proposition 3.1.14 implies that P and Q are $\mathbb{F}_q$-conjugate. This shows that ϕ is injective.

Let $\mathcal{P}$ be a place of degree m of $\mathbb{F}_q(\mathcal{X})$. By Theorem 1.7.2, there are exactly m places of degree 1 of $\mathbb{F}_q(\mathcal{X}) \cdot \mathbb{F}_{q^m}$ lying over $\mathcal{P}$. Let $\mathcal{N}$ be one of these m places and let R be its valuation ring. Then R is a discrete valuation ring with quotient field $\mathbb{F}_q(\mathcal{X}) \cdot \mathbb{F}_{q^m}$. It is easy to see that $R \cdot \overline{\mathbb{F}_q}$ is a discrete valuation ring with quotient field $\mathbb{F}_q(\mathcal{X}) \cdot \overline{\mathbb{F}_q} = \overline{\mathbb{F}_q}(\mathcal{X})$. Hence, by Theorem 3.1.12, there is a point P of $\mathcal{X}$ such that $\mathcal{O}_P(\mathcal{X}) = R \cdot \overline{\mathbb{F}_q}$. It is clear that ϕ maps the $\mathbb{F}_q$-closed point P containing P to $\mathcal{P}$. Hence, ϕ is surjective.

Furthermore, each place of $\mathbb{F}_q(\mathcal{X}) \cdot \mathbb{F}_{q^m}$ lying over $\mathcal{P}$ yields a point of $\mathcal{X}$ and these m points are $\mathbb{F}_q$-conjugate, that is, they belong to the same $\mathbb{F}_q$-closed point P. Conversely, any point P in P gives rise to a place of $\mathbb{F}_q(\mathcal{X}) \cdot \mathbb{F}_{q^m}$ lying over $\mathcal{P}$. This implies that the degree of P is equal to the degree of $\mathcal{P}$. $\qquad\square$

Remark 3.1.16. Let $\mathcal{X}$ be a nonsingular projective curve over $\mathbb{F}_q$. Then by Theorem 3.1.15, an $\mathbb{F}_q$-closed point of $\mathcal{X}$ corresponds to a uniquely determined place of $\mathbb{F}_q(\mathcal{X})$. Thus, a local parameter at this place of $\mathbb{F}_q(\mathcal{X})$ can be called a *local parameter* or *uniformizing parameter* at the corresponding $\mathbb{F}_q$-closed point.

Example 3.1.17. Consider the projective line $\mathcal{X} = \mathbf{P}^1(\overline{\mathbb{F}_q})$. It is a nonsingular projective curve over $\mathbb{F}_q$ and its $\mathbb{F}_q$-rational function field is the rational function field $\mathbb{F}_q(x)$ over $\mathbb{F}_q$ in the variable x. The places of $\mathbb{F}_q(x)$ have been determined in Section 1.5; they are the finite places corresponding to the monic irreducible polynomials over $\mathbb{F}_q$ and the infinite place. By Theorem 3.1.15, each finite place corresponds to an $\mathbb{F}_q$-closed point of $\mathcal{X}$ whose degree is equal to the degree of the corresponding monic irreducible polynomial, whereas the infinite place corresponds to an $\mathbb{F}_q$-rational point of $\mathcal{X}$. For an $\mathbb{F}_q$-closed point P of $\mathcal{X}$ corresponding to a finite place, the points in P are associated with the roots of the corresponding monic irreducible polynomial in $\overline{\mathbb{F}_q}$.

3.2 Maps Between Curves

We refer to Section 2.5 for background on morphisms and rational maps.

Lemma 3.2.1. A rational map from a variety to a curve is either constant or dominant.

Proof. Let ϕ be a rational map from a variety to a curve $\mathcal{Y}$. If ϕ is not constant, then we have at least two points in $\mathcal{W} := \phi(\mathrm{dom}(\phi))$. Thus, if $Q \in W$, then $\{Q\} \subsetneq \overline{W} \subseteq \mathcal{Y}$. Since the dimension of $\mathcal{Y}$ is equal to 1, we must have $\overline{W} = \mathcal{Y}$, that is, ϕ is dominant.   □

Definition 3.2.2. A rational map $\phi : \mathcal{X}_1/k \to \mathcal{X}_2/k$ between projective curves is *defined over k* if $\phi \circ \sigma = \sigma \circ \phi$ for all $\sigma \in \mathrm{Gal}(\overline{k}/k)$, that is, if $\phi(\sigma(P)) = \sigma(\phi(P))$ for all $P \in \mathrm{dom}(\phi)$ and all $\sigma \in \mathrm{Gal}(\overline{k}/k)$.

Theorem 3.2.3. Let $\phi : \mathcal{X}_1/k \to \mathcal{X}_2/k$ be a nonconstant rational map defined over k between projective curves. Then:

(i) The induced map

$$\phi^* : k(\mathcal{X}_2) \to k(\mathcal{X}_1), \qquad f \mapsto f \circ \phi,$$

between k-rational function fields is a homomorphism fixing k. Moreover, this homomorphism is independent of the choice of ϕ in its equivalence class.

(ii) The field extension $k(\mathcal{X}_1)/\phi^*(k(\mathcal{X}_2))$ is finite.

(iii) If a point P belongs to the domain $\mathrm{dom}(\phi)$ and $\phi(P) = Q$, then $\mathcal{O}_P(\mathcal{X}_1)$ dominates $\phi^*(\mathcal{O}_Q(\mathcal{X}_2))$. Conversely, if $\mathcal{O}_P(\mathcal{X}_1)$ dominates $\phi^*(\mathcal{O}_Q(\mathcal{X}_2))$ for points P of $\mathcal{X}_1$ and Q of $\mathcal{X}_2$, then P belongs to the domain $\mathrm{dom}(\phi)$ and $\phi(P) = Q$.

Proof.

(i) By Lemma 3.2.1, ϕ is dominant. Moreover, by Theorem 2.5.24,

$$f \in \overline{k}(\mathcal{X}_2) \mapsto f \circ \phi \in \overline{k}(\mathcal{X}_1)$$

is a homomorphism. Since ϕ is defined over k, it induces a homomorphism $\phi^* : k(\mathcal{X}_2) \to k(\mathcal{X}_1)$. It is clear that this homomorphism fixes k and is independent of the choice of ϕ in its equivalence class.

(ii) As both $k(\mathcal{X}_1)$ and $k(\mathcal{X}_2)$ are algebraic function fields of one variable with full constant field k and $\phi^*(k(\mathcal{X}_2))$ is a subfield of $k(\mathcal{X}_1)$, then by Definition 1.5.2, $k(\mathcal{X}_1)/\phi^*(k(\mathcal{X}_2))$ is a finite extension.

(iii) It is easy to verify the first statement. To show the converse, we let V and W be affine neighborhoods of P and Q, respectively, such that ϕ is a morphism from V to W. By (i) and Theorem 2.5.5, ϕ^* induces a homomorphism from $\overline{k}[W] = \overline{k}[x_1, \ldots, x_n]/I(W)$ to $\mathcal{O}_V$, which we call also ϕ^*. Now by the assumption that $\mathcal{O}_P(\mathcal{X}_1)$ dominates $\phi^*(\mathcal{O}_Q(\mathcal{X}_2))$, we have $g(\phi(P)) = (\phi^*g)(P) = 0$ for any $g \in \overline{k}[W]$ with $g(Q) = 0$. By taking g to be the residue class of $x_i - a_i$ in $\overline{k}[W]$ with $Q = (a_1, \ldots, a_n)$ and $i = 1, \ldots, n$, we get $\phi(P) = Q$. $\qquad\square$

The *degree* of a nonconstant rational map ϕ defined over k from a projective curve $\mathcal{X}_1/k$ to a projective curve $\mathcal{X}_2/k$ is defined by $[k(\mathcal{X}_1) : \phi^*(k(\mathcal{X}_2))]$.

Lemma 3.2.4. Let $\mathcal{X}_1/k$ and $\mathcal{X}_2/k$ be two projective curves. Then any homomorphism from $k(\mathcal{X}_2)$ into $k(\mathcal{X}_1)$ is induced by a unique dominant rational map defined over k from $\mathcal{X}_1/k$ to $\mathcal{X}_2/k$.

Proof. A homomorphism from $k(\mathcal{X}_2)$ into $k(\mathcal{X}_1)$ can be naturally extended to a homomorphism from $\overline{k}(\mathcal{X}_2)$ into $\overline{k}(\mathcal{X}_1)$. By Theorem 2.5.24, there is a unique dominant rational map giving this homomorphism, and it is easy to verify that this map is defined over k. $\qquad\square$

Lemma 3.2.5. Let $K \subseteq L$ be two algebraic function fields of one variable with the same full constant field. Then any ring R satisfying $K \subseteq R \subseteq L$ is a field.

Proof. Let $\alpha \in R$ with $\alpha \neq 0$. Then α is algebraic over K since L/K is an algebraic extension. Thus, α satisfies an equation

$$\alpha^n + a_1\alpha^{n-1} + \cdots + a_{n-1}\alpha + a_n = 0$$

for some $a_1, \ldots, a_n \in K$ and $a_n \neq 0$. Hence,

$$\alpha\left(\alpha^{n-1} + a_1\alpha^{n-2} + \cdots + a_{n-1}\right)\left(-a_n^{-1}\right) = 1.$$

This means that α^{-1} belongs to R, and so R is a field. $\qquad\square$

Proposition 3.2.6. Let $\phi : \mathcal{X}_1/k \to \mathcal{X}_2/k$ be a rational map defined over k between projective curves. Then the domain $\mathrm{dom}(\phi)$ contains every nonsingular point of $\mathcal{X}_1/k$. In particular, if $\mathcal{X}_1/k$ is a nonsingular projective curve, then ϕ is a morphism.

Proof. If ϕ is constant, then $\text{dom}(\phi) = \mathcal{X}_1$ and the result is true. Now assume that ϕ is not a constant map. By Theorem 3.2.3, ϕ induces a homomorphism $\phi^* : k(\mathcal{X}_2) \to k(\mathcal{X}_1)$. Let P be a nonsingular point of $\mathcal{X}_1$. Then by Lemma 3.1.11 and its proof and Theorem 3.2.3(iii), it suffices to show that $\overline{k}(\mathcal{X}_2) \not\subseteq \mathcal{O}_P(\mathcal{X}_1)$. Suppose $\overline{k}(\mathcal{X}_2) \subseteq \mathcal{O}_P(\mathcal{X}_1)$. Then $\mathcal{O}_P(\mathcal{X}_1)$ is a field by Lemma 3.2.5. This is a contradiction as the discrete valuation ring $\mathcal{O}_P(\mathcal{X}_1)$ cannot be a field because a field does not contain irreducible elements. $\qquad\square$

The following three results form the culmination points of this section.

Theorem 3.2.7. Two nonsingular projective curves over k are k-isomorphic if and only if their function fields are k-isomorphic.

Proof. The necessity is clear as an isomorphism is a birational map. Conversely, if the function fields are k-isomorphic, then the curves are birationally equivalent. Since both curves are nonsingular, it follows from Proposition 3.2.6 that the birational map between them is an isomorphism. $\qquad\square$

Theorem 3.2.8. Let $\mathcal{X}/k$ be a projective curve. Then there is a non-singular projective curve $\mathcal{Y}/k$ such that there exists a birational map ϕ from $\mathcal{Y}$ to $\mathcal{X}$. If φ is a birational map from a nonsingular projective curve $\mathcal{Y}'/k$ to $\mathcal{X}$, then there exists a unique isomorphism $\theta : \mathcal{Y} \to \mathcal{Y}'$ such that $\varphi \circ \theta = \phi$.

Proof. By Corollary 2.5.26, $\mathcal{X}$ is birationally equivalent to a plane curve $\mathcal{Z}$. By resolution of singularities (see [35, Chapter 7] and [51, Section 1.4]), we can find a nonsingular projective curve $\mathcal{Y}$ birationally equivalent to $\mathcal{Z}$.

For the second part, since $\mathcal{Y}$ and $\mathcal{Y}'$ are birationally equivalent, we have by Proposition 3.2.6 and Theorem 3.2.7 that $\overline{k}(\mathcal{Y})$ is isomorphic to $\overline{k}(\mathcal{Y}')$. Now uniqueness follows from Theorem 2.5.24. $\qquad\square$

Theorem 3.2.9. There is a one-to-one correspondence between k-isomorphism classes of nonsingular projective curves over k and

k-isomorphism classes of algebraic function fields of one variable with full constant field k, induced by

$$\chi : \mathcal{X}/k \rightarrow k(\mathcal{X}).$$

Proof. First we prove that χ is surjective on k-isomorphism classes. Let F be a given algebraic function field of one variable with full constant field k. Then there exists $z \in F$ transcendental over k such that $[F : k(z)] < \infty$. Thus, $F = k(z, y_1, \ldots, y_n)$ with $y_1, \ldots, y_n \in F$ algebraic over $k(z)$. Let I be the kernel of the natural homomorphism from the polynomial ring $\overline{k}[Z, Y_1, \ldots, Y_n]$ in the variables $Z, Y_1, \ldots, Y_n$ to the ring E formed by the polynomial expressions in $z, y_1, \ldots, y_n$ with coefficients from $\overline{k}$. Then I is a nonzero prime ideal of $\overline{k}[Z, Y_1, \ldots, Y_n]$. Let $\mathcal{Y}$ be the zero set of I. Then $I(\mathcal{Y}) = I$ by Remark 2.2.9 and $\mathcal{Y}$ is an affine variety by Theorem 2.3.8. Moreover, $\mathcal{Y}$ is defined over k. By the definition of I, the coordinate ring $\overline{k}[\mathcal{Y}]$ of $\mathcal{Y}$ is isomorphic to E. Hence, by Theorem 2.4.13(iv), $\overline{k}(\mathcal{Y}) \simeq \overline{k}(z, y_1, \ldots, y_n)$ and $\mathcal{Y}$ is a curve. The projective closure of $\mathcal{Y}/k$ (compare with Proposition 2.3.12) again has $\overline{k}(\mathcal{Y})$ as its function field, according to Remark 2.4.14(ii). Now we apply Theorem 3.2.8 and we obtain a nonsingular projective curve $\mathcal{X}$, which is birationally equivalent to the projective closure of $\mathcal{Y}/k$. By Corollary 2.5.25 we get $\overline{k}(\mathcal{X}) \simeq \overline{k}(\mathcal{Y}) \simeq \overline{k}(z, y_1, \ldots, y_n)$. Remark 2.4.19(ii) implies then that $k(\mathcal{X}) \simeq k(z, y_1, \ldots, y_n) = F$. Hence, χ is indeed surjective on k-isomorphism classes. By Theorem 3.2.7, χ is injective on k-isomorphism classes. $\square$

3.3 Divisors

We recall from Section 3.1 that for a given nonsingular projective curve over k, its k-rational function field is an algebraic function field of one variable with full constant field k. Conversely, Theorem 3.2.9 shows that given an algebraic function field F/k of one variable with full constant field k, there exists a nonsingular projective curve $\mathcal{X}/k$ such that $F \simeq k(\mathcal{X})$. Furthermore, results on curves can be used for function fields with a proper interpretation.

We now start from an algebraic function field F/k of one variable with full constant field k. In view of the preceding remarks, the theory of divisors of F that we discuss in the following can be developed in an equivalent fashion in the language of nonsingular projective curves. We write $\mathbf{P}_F$ for the set of places of F. The *divisor group* of F/k, denoted by $\mathrm{Div}(F/k)$ or $\mathrm{Div}(F)$, is the free abelian group generated by the places of F, that is, a *divisor* in $\mathrm{Div}(F)$ (also called a divisor of F) is a formal sum

$$\sum_{P \in \mathbf{P}_F} n_P P$$

with coefficients $n_P \in \mathbb{Z}$ and $n_P = 0$ for all but finitely many $P \in \mathbf{P}_F$, and two divisors of F are added by adding the corresponding coefficients. A divisor $D = \sum_{P \in \mathbf{P}_F} n_P P$ of F is called *positive* (or *effective*), written as $D \geq 0$, if $n_P \geq 0$ for all $P \in \mathbf{P}_F$. For two divisors D and G of F with $D - G$ being positive, we write $D \geq G$ or $G \leq D$.

For $D = \sum_{P \in \mathbf{P}_F} n_P P \in \mathrm{Div}(F)$, it is often convenient to put $v_P(D) := n_P$ for $P \in \mathbf{P}_F$. The *support* $\mathrm{supp}(D)$ of D is the finite set given by

$$\mathrm{supp}(D) := \{P \in \mathbf{P}_F : v_P(D) \neq 0\}.$$

The *degree* $\deg(D)$ of the divisor $D = \sum_{P \in \mathbf{P}_F} n_P P$ is defined by

$$\deg(D) = \deg\left(\sum_{P \in \mathbf{P}_F} n_P P\right) = \sum_{P \in \mathbf{P}_F} n_P \deg(P).$$

It is clear that $\deg$ defines a group homomorphism from $\mathrm{Div}(F)$ to $\mathbb{Z}$. The kernel of this homomorphism is a subgroup of $\mathrm{Div}(F)$, denoted by $\mathrm{Div}^0(F)$, that is,

$$\mathrm{Div}^0(F) := \{D \in \mathrm{Div}(F) : \deg(D) = 0\}.$$

Proposition 3.3.1. For any $x \in F \setminus k$, we have

$$\sum_{\substack{P \in \mathbf{P}_F \\ v_P(x) > 0}} v_P(x) \deg(P) \leq [F : k(x)].$$

Proof. If $x \in F \setminus k$, then $x \notin \bar{k}$ since k is the full constant field of F, and so $d := [F : k(x)] < \infty$. Suppose that

$$\sum_{\substack{P \in \mathbf{P}_F \\ v_P(x) > 0}} v_P(x) \deg(P) > d.$$

Then there exist distinct places $P_1, \ldots, P_r$ of F such that $n_i := v_{P_i}(x) > 0$ for $1 \leq i \leq r$ and $\sum_{i=1}^{r} n_i \deg(P_i) > d$. Put $\mathcal{O} = \bigcap_{i=1}^{r} \mathcal{O}_{P_i}$ and note that $x \in \mathcal{O}$. By Theorem 1.5.18, for each $i = 1, \ldots, r$ we can choose an element $t_i \in F$ such that $v_{P_i}(t_i) = -1$ and $v_{P_h}(t_i) = 0$ for all h with $1 \leq h \leq r$ and $h \neq i$. Furthermore, there exist elements $u_{im} \in \mathcal{O}, 1 \leq m \leq \deg(P_i)$, such that the set $\{u_{im}(P_i)\}_{1 \leq m \leq \deg(P_i)}$ of residue classes forms a k-basis of the residue class field of P_i. In order to arrive at the desired contradiction, it suffices to show that the elements $u_{im} t_i^j \in F$, $1 \leq m \leq \deg(P_i)$, $1 \leq j \leq n_i$, $1 \leq i \leq r$, are linearly independent over $k(x)$. If there were a nontrivial linear dependence relation, then by clearing denominators and afterwards dividing by a suitable power of x, we can put it in the form

$$\sum_{i=1}^{r} \sum_{j=1}^{n_i} f_{ij} t_i^j + x \sum_{i=1}^{r} \sum_{j=1}^{n_i} g_{ij} t_i^j = 0,$$

where all $f_{ij}, g_{ij} \in \mathcal{O}$, either $f_{ij} = 0$ or $v_{P_i}(f_{ij}) = 0$, and the latter case occurs for at least one pair (i, j). Now fix a subscript ℓ such that $v_{P_\ell}(f_{\ell j}) = 0$ for some j with $1 \leq j \leq n_\ell$. Then

$$v_{P_\ell}\left(\sum_{i=1}^{r} \sum_{j=1}^{n_i} f_{ij} t_i^j\right) < 0$$

and

$$v_{P_\ell}\left(x \sum_{i=1}^{r} \sum_{j=1}^{n_i} g_{ij} t_i^j\right) \geq 0,$$

a contradiction. □

We recall a terminology introduced in Section 1.5, according to which we say for a place P of F and an element $x \in F^*$ that P is a zero of x if $v_P(x) > 0$ and that P is a pole of x if $v_P(x) < 0$.

Corollary 3.3.2. Every element of F^* has only finitely many zeros and finitely many poles.

Proof. If x is a nonzero element in k, then x has neither zeros nor poles. If $x \in F \setminus k$, then x is transcendental over k, and so by Proposition 3.3.1 the number of zeros of x is at most $[F : k(x)]$, which is finite. The same argument shows that x^{-1} has only finitely many zeros, that is, x has only finitely many poles. □

In view of Corollary 3.3.2, it makes sense to introduce the following divisors of F associated with an element $x \in F^*$. Let $\mathcal{N}(x)$ be the set of zeros and $\mathcal{P}(x)$ the set of poles of x. Note that $\mathcal{N}(x)$ and $\mathcal{P}(x)$ are finite sets by Corollary 3.3.2. We define the *zero divisor* $(x)_0$ of x by

$$(x)_0 = \sum_{P \in \mathcal{N}(x)} v_P(x) P$$

and the *pole divisor* $(x)_\infty$ of x by

$$(x)_\infty = \sum_{P \in \mathcal{P}(x)} (-v_P(x)) P.$$

Note that both $(x)_0$ and $(x)_\infty$ are positive divisors. Finally, we define the *principal divisor* $\operatorname{div}(x)$ of x by

$$\operatorname{div}(x) = (x)_0 - (x)_\infty = \sum_{P \in \mathbf{P}_F} v_P(x) P.$$

It is clear that $\operatorname{div} : x \in F^* \mapsto \operatorname{div}(x) \in \operatorname{Div}(F)$ is a group homomorphism. The following result determines the kernel of div.

Proposition 3.3.3. Every element of $F \setminus k$ has at least one zero and at least one pole. In particular, the kernel of div is k^*.

Proof. Suppose that $x \in F \setminus k$ has no pole. Then $v_P(x) \geq 0$ for all points P of the nonsingular projective curve $\mathcal{X}/k$ corresponding to F by Theorem 3.2.9, and so the values $x(P) \in \bar{k}$ make sense. Consider the map

$$\phi_x : \mathcal{X}/k \to \mathbf{P}^1(\bar{k}), \quad P \mapsto [x(P), 1].$$

Then ϕ_x is a morphism. Moreover, ϕ_x is not surjective as the point $[1, 0]$ is not in the image Y of ϕ_x. On the other hand, Y is an irreducible closed subset of $\mathbf{P}^1(\bar{k})$, and so Y is a singleton. This contradiction shows that x has at least one pole. Since $1/x$ has at least one pole as well, x has at least one zero.

It is clear that k^* is contained in the kernel of div. Conversely, if $x \in F^*$ is such that $\text{div}(x) = 0 \in \text{Div}(F)$, then x has no zeros and poles, and hence x is an element of k^* by what we have already shown. □

3.4 Riemann-Roch Spaces

For a divisor D of F/k, we define the *Riemann-Roch space*

$$\mathcal{L}(D) := \{x \in F^* : \text{div}(x) + D \geq 0\} \cup \{0\}.$$

Then it is easy to verify that $\mathcal{L}(D)$ is a vector space over k. We denote by $\ell(D)$ the dimension of $\mathcal{L}(D)$ as a vector space over k. The following proposition shows some basic properties of $\mathcal{L}(D)$, in particular, that $\ell(D)$ is finite for every divisor D of F.

Proposition 3.4.1. Let D and G be divisors of F/k. Then:

(i) if $D \leq G$, then $\mathcal{L}(D)$ is a subspace of $\mathcal{L}(G)$ and

$$\dim_k(\mathcal{L}(G)/\mathcal{L}(D)) \leq \deg(G) - \deg(D);$$

(ii) $\mathcal{L}(0) = k$;

(iii) $\ell(D) \geq 1$ if $D \geq 0$;

(iv) $\ell(D)$ is finite for all D;

(v) if $D = G + \text{div}(x)$ for some nonzero $x \in F$, then $\ell(D) = \ell(G)$.

Proof.

(i) If $x \in \mathcal{L}(D) \setminus \{0\}$, then by definition $\text{div}(x) + D \geq 0$. Thus, $\text{div}(x) + G = \text{div}(x) + D + (G - D) \geq 0$. Hence, $x \in \mathcal{L}(G)$ and $\mathcal{L}(D) \subseteq \mathcal{L}(G)$.

For the second assertion, it suffices to consider the case where $G = D + P$ for some place P of F, as the general case follows then by iteration. Choose $u \in F$ with $v_P(u) = v_P(G) = v_P(D) + 1$. For any $x \in \mathcal{L}(G)$ we have $v_P(x) \geq -v_P(G) = -v_P(u)$; hence, $ux \in \mathcal{O}_P$. Thus, we get a k-linear map ψ from $\mathcal{L}(G)$ to the residue class field $F_P = \mathcal{O}_P/M_P$ of P defined by $\psi(x) = (ux)(P)$ for all $x \in \mathcal{L}(G)$. The kernel of ψ is $\mathcal{L}(D)$, and so

$$\dim_k(\mathcal{L}(G)/\mathcal{L}(D)) \leq \dim_k(F_P) = \deg(P) = \deg(G) - \deg(D).$$

(ii) An element of k^* has neither zeros nor poles, hence $\mathcal{L}(0)$ contains k. On the other hand, by Proposition 3.3.3 any $x \in F \setminus k$ has at least one pole, that is, $x \notin \mathcal{L}(0)$. Thus, $\mathcal{L}(0) = k$.

(iii) By (i) and (ii) we know that $k = \mathcal{L}(0)$ is a subspace of $\mathcal{L}(D)$. The desired result follows.

(iv) If $D \geq 0$, then we can apply (i) and (ii) to obtain

$$\ell(D) = \dim_k(\mathcal{L}(D)/\mathcal{L}(0)) + 1 \leq \deg(D) + 1, \qquad (3.1)$$

and so $\ell(D) < \infty$. If D is arbitrary, then $D \leq H$ for some divisor $H \geq 0$, and so (i) yields $\mathcal{L}(D) \subseteq \mathcal{L}(H)$. But $\ell(H) < \infty$ by what we have already shown, hence, $\ell(D) < \infty$.

(v) It is easy to verify that $x\mathcal{L}(D) = \mathcal{L}(G)$. $\square$

Theorem 3.4.2. For any $x \in F \setminus k$, we have

$$\deg((x)_0) = [F : k(x)].$$

Proof. Set $n = [F : k(x)]$. By Proposition 3.3.1 we have $\deg((x)_0) \leq n$. It remains to show that $\deg((x)_0) \geq n$. Choose a basis $y_1, \ldots, y_n$ of the extension $F/k(x)$. We introduce the positive divisor $C = \sum_{j=1}^{n}(y_j)_\infty$ of F. For any integer $m \geq 1$, observe that $x^{-i}y_j \in \mathcal{L}(m(x)_0 + C)$ for all $0 \leq i \leq m, 1 \leq j \leq n$, and that the elements $x^{-i}y_j$, $0 \leq i \leq m, 1 \leq j \leq n$, are linearly independent over k. This yields

$$\ell(m(x)_0 + C) \geq n(m + 1) \qquad (3.2)$$

for all $m \geq 1$. On the other hand, (3.1) shows that

$$\ell(m(x)_0 + C) \leq m \deg((x)_0) + \deg(C) + 1. \qquad (3.3)$$

Combining (3.2) and (3.3), we get

$$m(\deg((x)_0) - n) \geq n - \deg(C) - 1$$

for all $m \geq 1$. The above inequality implies that $\deg((x)_0) \geq n$ since the right-hand side is independent of m.   □

Corollary 3.4.3. For any nonzero $x \in F$, we have $\deg(\mathrm{div}(x)) = 0$, that is, $\deg((x)_0) = \deg((x)_\infty)$.

Proof. The result is trivial for $x \in k^*$. For $x \in F \setminus k$, we note the fact that $(1/x)_0 = (x)_\infty$. By Theorem 3.4.2 we obtain

$$\deg((x)_0) = [F : k(x)] = [F : k(1/x)] = \deg((1/x)_0) = \deg((x)_\infty),$$

which is the desired result.   □

Corollary 3.4.4. We have $\ell(D) = 0$ whenever $\deg(D) < 0$.

Proof. Suppose $\ell(D) > 0$. Then there exists a nonzero $x \in \mathcal{L}(D)$, that is, we have $\mathrm{div}(x) + D \geq 0$. Hence, by Corollary 3.4.3, $0 \leq \deg(\mathrm{div}(x) + D) = \deg(D)$. This is a contradiction.   □

It is easy to verify that all principal divisors form a subgroup of $\mathrm{Div}(F)$, denoted by $\mathrm{Princ}(F)$. The quotient group $\mathrm{Div}(F)/\mathrm{Princ}(F)$ is called the *divisor class group* of F. Two divisors D and G belonging to the same residue class of $\mathrm{Div}(F)/\mathrm{Princ}(F)$ are said to be *equivalent*, written as $D \sim G$. If $D \sim G$, then $\ell(D) = \ell(G)$ by Proposition 3.4.1(v) and $\deg(D) = \deg(G)$ by Corollary 3.4.3. Furthermore, Corollary 3.4.3 shows that $\mathrm{Princ}(F)$ is also a subgroup of $\mathrm{Div}^0(F)$, the group of divisors of F of degree 0.

We conclude this section with a useful characterization of rational function fields.

Proposition 3.4.5. The following are equivalent:

  (i) F is the rational function field over k;
  (ii) F is the k-rational function field of a curve that is k-isomorphic to $\mathbf{P}^1(\overline{k})$;
  (iii) there exists an element $x \in F^*$ with $\deg((x)_0) = 1$;
  (iv) there exists a rational place P of F with $\ell(P) = 2$.

Proof. (iv) says that there exists an element $x \in \mathcal{L}(P) \setminus k$ such that $\mathrm{div}(x) + P \geq 0$. Thus, $(x)_\infty = P$ and hence by Theorem 3.4.2 and Corollary 3.4.3, $[F : k(x)] = \deg((x)_0) = \deg((x)_\infty) = 1$, so that $F = k(x)$. The remaining assertions are either obvious or follow easily from Theorems 3.2.7 and 3.4.2; compare also with Example 3.1.17.

$\square$

3.5 Riemann's Theorem and Genus

In this section, we are preparing the ground for one of the fundamental results on curves and function fields, the Riemann-Roch theorem. In this and the following section, a function field F is meant to be an algebraic function field of one variable. As usual, the full constant field of F is denoted by k and assumed to be perfect.

Theorem 3.5.1 (Riemann's Theorem). For any function field F, there exists a nonnegative integer g depending only on F such that

$$\ell(D) \geq \deg(D) + 1 - g \tag{3.4}$$

for every divisor D of F.

Proof. For each divisor D of F, we set $s(D) = \deg(D) + 1 - \ell(D)$. The task is to find a nonnegative $g \in \mathbb{Z}$ such that $s(D) \leq g$ for all D.

Let x be an element in $F \setminus k$ and set $A = (x)_0$. Then as in (3.2), we have a positive divisor C of F such that for all integers $m \geq 1$ we get

$$\ell(mA + C) \geq \deg(A)(m + 1).$$

By Proposition 3.4.1(i) we have

$$\ell(mA + C) \leq \ell(mA) + \deg(C).$$

Hence,

$$\ell(mA) \geq \deg(A)(m + 1) - \deg(C),$$

that is,

$$s(mA) = \deg(mA) + 1 - \ell(mA) \leq \deg(C) - \deg(A) + 1$$

for all m. Thus, $s(mA)$ is bounded by a constant, which is independent of m. Set

$$g = \max \{s(mA) : m \geq 1\}. \tag{3.5}$$

Now let D be an arbitrary divisor of F. Choose a positive divisor $G \geq D$ and observe that by Proposition 3.4.1(i) we get

$$\ell(mA - G) \geq \ell(mA) - \deg(G)$$

for all m. Then

$$\ell(mA) = \deg(mA) + 1 - s(mA) \geq \deg(mA) + 1 - g$$

by (3.5), and so

$$\ell(mA - G) \geq \deg(mA) + 1 - g - \deg(G) \geq 1$$

for a sufficiently large m. Thus, for this m there exists a nonzero $z \in \mathcal{L}(mA - G)$. With $H := G - \operatorname{div}(z)$ we get $H \sim G$ and $H \leq mA$. Now $s(D) \leq s(G)$ by Proposition 3.4.1(i). Since $\deg(G) = \deg(H)$ and $\ell(G) = \ell(H)$, it follows that $s(D) \leq s(H) \leq s(mA) \leq g$, where we again used Proposition 3.4.1(i) in the second inequality and (3.5) in the last inequality. Thus, (3.4) is established. Finally, noting that $s(0) = 0$ by Proposition 3.4.1(ii), we obtain $g \geq 0$. $\square$

Corollary 3.5.2. If $\ell(D) = \deg(D) + 1 - g$ and $G \geq D$, then $\ell(G) = \deg(G) + 1 - g$.

Proof. We have $\ell(G) \leq \deg(G) + 1 - g$ by Proposition 3.4.1(i) and $\ell(G) \geq \deg(G) + 1 - g$ by Theorem 3.5.1. $\square$

Corollary 3.5.3. There exists an integer r depending only on F such that

$$\ell(D) = \deg(D) + 1 - g$$

for all divisors D of F with $\deg(D) \geq r$.

Proof. By the proof of Theorem 3.5.1, we can find a divisor D_0 of F such that $\ell(D_0) = \deg(D_0) + 1 - g$ (see (3.5)). Let $r = \deg(D_0) + g$. Then by Theorem 3.5.1, for any divisor D of F with $\deg(D) \geq r$, we have $\ell(D - D_0) \geq \deg(D - D_0) + 1 - g \geq 1$, that is, there exists an $x \in F^*$ such that $D + \operatorname{div}(x) \geq D_0$. Then Corollary 3.5.2 yields

$$\ell(D) = \ell(D + \operatorname{div}(x)) = \deg(D + \operatorname{div}(x)) + 1 - g = \deg(D) + 1 - g,$$

which is the desired result. □

In view of Corollary 3.5.3, the integer g is uniquely determined by the function field F. This leads to the following important definition.

Definition 3.5.4. The nonnegative integer g determined by Corollary 3.5.3 is called the *genus* of the function field F. The *genus* of a nonsingular projective curve $\mathcal{X}$ over k is defined to be the genus of its k-rational function field $k(\mathcal{X})$.

Example 3.5.5. Let $F = k(x)$ be the rational function field over k. For the infinite place P_∞ of F and any integer $n \geq 0$, the Riemann-Roch space $\mathcal{L}(n P_\infty)$ consists of all polynomials from $k[x]$ of degree at most n. Hence, $\ell(n P_\infty) = n + 1$ for all $n \geq 0$ and F has genus $g = 0$ by Corollary 3.5.3. Conversely, if a function field F has genus 0 and at least one rational place P, then $\ell(P) \leq 2$ by (3.1) and $\ell(P) \geq 2$ by Theorem 3.5.1, and so $\ell(P) = 2$. Hence, F is a rational function field by Proposition 3.4.5.

3.6 The Riemann-Roch Theorem

We need further preparations for the proof of the Riemann-Roch theorem, in particular, the definition of the following concept.

Definition 3.6.1. An *adèle* of a function field F/k is a map $\alpha : \mathbf{P}_F \to F$ defined by $P \mapsto \alpha_P$ such that $\alpha_P \in \mathcal{O}_P$ for all but finitely many places P of F. We may regard an adèle as an element of the direct product $\prod_{P \in \mathbf{P}_F} F$.

The associated space $\mathcal{A}_F$ that consists of all adèles of F/k is called the *adèle space* of F and is obviously a vector space over k, with the algebraic structure inherited from $\prod_{P \in \mathbf{P}_F} F$.

Given an element $x \in F$, the *principal adèle* of x is the adèle whose components are all equal to x; this definition agrees with the condition imposed on adèles in view of Corollary 3.3.2. This construction yields an embedding $F \hookrightarrow \mathcal{A}_F$. The valuations v_P of F extend naturally to $\mathcal{A}_F$ by setting $v_P(\alpha) = v_P(\alpha_P)$. By Definition 3.6.1, $v_P(\alpha) \geq 0$ for all but finitely many $P \in \mathbf{P}_F$.

Definition 3.6.2. For any divisor $D \in \mathrm{Div}(F)$, we set

$$\mathcal{A}_F(D) := \{\alpha \in \mathcal{A}_F : v_P(\alpha) + v_P(D) \geq 0 \text{ for all } P \in \mathbf{P}_F\}.$$

It is evident that $\mathcal{A}_F(D)$ is a k-linear subspace of $\mathcal{A}_F$. The following two lemmas provide information on the dimension of associated vector spaces over k.

Lemma 3.6.3. Let $D \leq G$ be two divisors of F/k. Then $\mathcal{A}_F(D) \subseteq \mathcal{A}_F(G)$ and

$$\dim_k(\mathcal{A}_F(G)/\mathcal{A}_F(D)) = \deg(G) - \deg(D).$$

Proof. The first assertion of the lemma is trivial. The proof of the second claim is somewhat similar to that of Proposition 3.4.1(i). Note first that it suffices to consider the case where $G = D + P$ for some $P \in \mathbf{P}_F$. Pick $u \in F$ with $v_P(u) = v_P(G) = v_P(D) + 1$ and consider the k-linear map $\theta : \mathcal{A}_F(G) \to F_P$ defined by $\alpha \mapsto (u\alpha_P)(P)$. We can easily see that θ is surjective with kernel $\mathcal{A}_F(D)$, and so

$$\dim_k(\mathcal{A}_F(G)/\mathcal{A}_F(D)) = \dim_k(F_P) = \deg(P) = \deg(G) - \deg(D).$$

This implies the desired result. $\qquad\square$

Lemma 3.6.4. Let $D \leq G$ be two divisors of F/k. Then

$$\dim_k((\mathcal{A}_F(G) + F)/(\mathcal{A}_F(D) + F)) = (\deg(G) - \ell(G)) - (\deg(D) - \ell(D)).$$

Proof. We have the exact sequence of k-linear maps

$$0 \to \mathcal{L}(G)/\mathcal{L}(D) \to \mathcal{A}_F(G)/\mathcal{A}_F(D) \to (\mathcal{A}_F(G) + F)/(\mathcal{A}_F(D) + F) \to 0.$$

Let σ_1 and σ_2 denote the second and the third map, respectively, in the above sequence. They are defined in the obvious manner. The only nontrivial part in the proof of exactness is the verification that the kernel of σ_2 is contained in the image of σ_1. Indeed, let $\alpha \in \mathcal{A}_F(G)$ with $\sigma_2(\alpha + \mathcal{A}_F(D)) = 0$. Then $\alpha \in \mathcal{A}_F(D) + F$; hence, there is some $x \in F$ such that $\alpha - x \in \mathcal{A}_F(D)$. From $\mathcal{A}_F(D) \subseteq \mathcal{A}_F(G)$ we conclude that $x \in \mathcal{A}_F(G) \cap F = \mathcal{L}(G)$. Thus, $\alpha + \mathcal{A}_F(D) = x + \mathcal{A}_F(D) = \sigma_1(x + \mathcal{L}(D))$ lies in the image of σ_1. From the exactness of the above sequence, we obtain that $\dim_k((\mathcal{A}_F(G) + F)/(\mathcal{A}_F(D) + F)) = \dim_k(\mathcal{A}_F(G)/\mathcal{A}_F(D)) - \dim_k(\mathcal{L}(G)/\mathcal{L}(D)) = (\deg(G) - \ell(G)) - (\deg(D) - \ell(D))$, where we used Lemma 3.6.3 in the second step. $\qquad\square$

Lemma 3.6.5. If F has genus g and D is a divisor of F with $\ell(D) = \deg(D) + 1 - g$, then $\mathcal{A}_F = \mathcal{A}_F(D) + F$.

Proof. Let $\alpha \in \mathcal{A}_F$ be given. By Definitions 3.6.1 and 3.6.2, we can find a divisor $G \geq D$ such that $\alpha \in \mathcal{A}_F(G)$. By Lemma 3.6.4 and Corollary 3.5.2, $\dim_k((\mathcal{A}_F(G) + F)/(\mathcal{A}_F(D) + F)) = (\deg(G) - \ell(G)) - (\deg(D) - \ell(D)) = (g - 1) - (g - 1) = 0$. This implies $\mathcal{A}_F(D) + F = \mathcal{A}_F(G) + F$ and since $\alpha \in \mathcal{A}_F(G)$, it follows that $\alpha \in \mathcal{A}_F(D) + F$, as desired. $\qquad\square$

We can now prove a preliminary version of the Riemann-Roch theorem.

Theorem 3.6.6. If F/k is a function field of genus g, then for any divisor D of F we have

$$\ell(D) = \deg(D) + 1 - g + \dim_k(\mathcal{A}_F/(\mathcal{A}_F(D) + F)).$$

Proof. Given D, we use Corollary 3.5.3 to obtain a divisor $D_0 \geq D$ such that $\ell(D_0) = \deg(D_0) + 1 - g$. By Lemma 3.6.5 we have $\mathcal{A}_F = \mathcal{A}_F(D_0) + F$. Applying Lemma 3.6.4, we obtain $\dim_k(\mathcal{A}_F/(\mathcal{A}_F(D) + F)) = \dim_k((\mathcal{A}_F(D_0) + F)/(\mathcal{A}_F(D) + F)) = (\deg(D_0) - \ell(D_0)) - (\deg(D) - \ell(D)) = g - 1 + \ell(D) - \deg(D)$, which is the desired result. $\qquad\square$

A further tool that we need is the theory of Weil differentials. Our treatment here is similar to that in Stichtenoth [117].

Definition 3.6.7. A *Weil differential* of a function field F/k is a k-linear map $\omega : \mathcal{A}_F \to k$ such that $\omega|_{\mathcal{A}_F(D)+F} \equiv 0$ for some divisor

$D \in \mathrm{Div}(F)$. Let Ω_F be the set of all Weil differentials of F and moreover define $\Omega_F(D)$ to be the collection of all Weil differentials of F that vanish on $\mathcal{A}_F(D) + F$.

We may regard $\Omega_F(D)$ as a vector space over k. In analogy with adèles, $\Omega_F(D)$ is a k-linear subspace of Ω_F.

Theorem 3.6.6 can be restated as

$$i(D) := \ell(D) - \deg(D) - 1 + g = \dim_k(\mathcal{A}_F/(\mathcal{A}_F(D) + F)) \qquad (3.6)$$

for any divisor D of F. The following result provides a second interpretation of the quantity $i(D)$.

Lemma 3.6.8. For any divisor D of F/k, we have

$$\dim_k(\Omega_F(D)) = i(D).$$

Proof. The space $\Omega_F(D)$ is canonically isomorphic to the space of k-linear forms on $\mathcal{A}_F/(\mathcal{A}_F(D) + F)$. Since $\mathcal{A}_F/(\mathcal{A}_F(D) + F)$ has finite dimension $i(D)$ by (3.6), we obtain the desired result. $\qquad\square$

Remark 3.6.9. A simple consequence of Lemma 3.6.8 is that $\Omega_F \neq \{0\}$. Indeed, pick a divisor D of F of degree less than -1. Then $\dim_k(\Omega_F(D)) = i(D) = \ell(D) - \deg(D) - 1 + g \geq 1$; hence, $\Omega_F(D) \neq \{0\}$ and a fortiori $\Omega_F \neq \{0\}$.

For any element $x \in F$ and any Weil differential ω on the space $\mathcal{A}_F$ of all adèles of F/k, we define $x\omega : \mathcal{A}_F \to k$ by letting $(x\omega)(\alpha) = \omega(x\alpha)$ for all $\alpha \in \mathcal{A}_F$. We see that $x\omega$ is again a Weil differential of F; in fact, if ω vanishes on $\mathcal{A}_F(D) + F$, then $x\omega$ vanishes on $\mathcal{A}_F(D + \mathrm{div}(x)) + F$. Clearly, this construction extends the set of scalars of Ω_F as a vector space from k to F. We now calculate the dimension of this new vector space.

Theorem 3.6.10. We have $\dim_F(\Omega_F) = 1$.

Proof. Choose a nonzero Weil differential $\omega_1 \in \Omega_F$; this is possible by Remark 3.6.9. We have to show that ω_1 generates the whole space or, equivalently, given any $\omega_2 \in \Omega_F$, there is $z \in F$ such that $\omega_2 = z\omega_1$. We can assume that $\omega_2 \neq 0$. Choose $D_1, D_2 \in \mathrm{Div}(F)$ such that $\omega_1 \in \Omega_F(D_1)$ and $\omega_2 \in \Omega_F(D_2)$. For a fixed divisor G of F and

$j = 1, 2$, consider the injective k-linear maps $\phi_j : \mathcal{L}(D_j + G) \to \Omega_F(-G)$ defined by $x \mapsto x\omega_j$. We claim that for an appropriate choice of the divisor G we have

$$\phi_1(\mathcal{L}(D_1 + G)) \cap \phi_2(\mathcal{L}(D_2 + G)) \neq \{0\}. \tag{3.7}$$

Let $G \neq 0$ be a positive divisor of sufficiently large degree such that

$$\ell(D_j + G) = \deg(D_j + G) + 1 - g$$

for $j = 1, 2$, which is possible by Corollary 3.5.3, and set $U_j = \phi_j(\mathcal{L}(D_j + G)) \subseteq \Omega_F(-G)$. Since $\dim_k(\Omega_F(-G)) = i(-G) = \deg(G) - 1 + g$ by Lemma 3.6.8 and Corollary 3.4.4, we obtain $\dim_k(U_1) + \dim_k(U_2) - \dim_k(\Omega_F(-G)) = \deg(D_1 + G) + 1 - g + \deg(D_2 + G) + 1 - g - \deg(G) + 1 - g = \deg(G) + \deg(D_1) + \deg(D_2) + 3(1 - g)$. Therefore,

$$\dim_k(U_1 \cap U_2) \geq \dim_k(U_1) + \dim_k(U_2) - \dim_k(\Omega_F(-G)) > 0$$

if the degree of G is sufficiently large. This establishes the claim (3.7). Now pick $x_1 \in \mathcal{L}(D_1 + G)$ and $x_2 \in \mathcal{L}(D_2 + G)$ such that $x_1\omega_1 = x_2\omega_2 \neq 0$; then $\omega_2 = (x_1 x_2^{-1})\omega_1$ as desired.   □

We will now proceed to attach a natural divisor of F to any nonzero Weil differential of F. Before that, we have to prove the following lemma.

Lemma 3.6.11. Let ω be a nonzero Weil differential of F. Then there is a uniquely determined divisor $W \in M(\omega) := \{D \in \mathrm{Div}(F) : \omega|_{A_F(D)+F} = 0\}$ such that $D \leq W$ for all $D \in M(\omega)$.

Proof. By Corollary 3.5.3 there exists an integer r depending only on F such that $i(D) = 0$ for all $D \in \mathrm{Div}(F)$ with $\deg(D) \geq r$. Since $\dim_k(A_F/(A_F(D) + F)) = i(D)$ by (3.6), we have $\deg(D) < r$ for all $D \in M(\omega)$. Thus, there exists a divisor $W \in M(\omega)$ of maximal degree.

We show that W satisfies the property in the lemma. If not, then there is a divisor $D_0 \in M(\omega)$ such that $v_Q(D_0) > v_Q(W)$ for some $Q \in \mathbf{P}_F$. We claim that then $W + Q \in M(\omega)$, which will be a contradiction to the choice of W. Indeed, consider an adèle $\alpha = (\alpha_P)_{P \in \mathbf{P}_F} \in \mathcal{A}_F(W + Q)$. We write $\alpha = \alpha' + \alpha''$ with

$$\alpha'_P = (1 - \delta_{PQ})\alpha_P, \quad \alpha''_P = \delta_{PQ}\alpha_P \quad \text{for all } P \in \mathbf{P}_F,$$

where $\delta_{PQ} = 1$ if $P = Q$ and $\delta_{PQ} = 0$ if $P \neq Q$. Then $\alpha' \in \mathcal{A}_F(W)$ and $\alpha'' \in \mathcal{A}_F(D_0)$, hence, $\omega(\alpha + x) = \omega(\alpha') + \omega(\alpha'' + x) = 0$ for all $x \in F$. Therefore, ω vanishes on $\mathcal{A}_F(W + Q) + F$, and our claim is true. This completes the proof of the lemma. $\square$

In view of the above lemma, the following definition is meaningful.

Definition 3.6.12. The divisor (ω) of a nonzero Weil differential ω of F is the unique divisor W of F such that: (i) ω vanishes on $\mathcal{A}_F(W) + F$; (ii) if ω vanishes on $\mathcal{A}_F(D) + F$, then $D \leq W = (\omega)$. A divisor (ω) for some nonzero Weil differential ω of F is called a *canonical divisor* of F.

Note now that $\Omega_F(D)$ introduced in Definition 3.6.7 can also be described by

$$\Omega_F(D) = \{\omega \in \Omega_F : \omega = 0 \text{ or } (\omega) \geq D\}.$$

Lemma 3.6.13. For any nonzero element $x \in F$ and any nonzero Weil differential $\omega \in \Omega_F$, we have $(x\omega) = \text{div}(x) + (\omega)$.

Proof. If ω vanishes on $\mathcal{A}_F(D) + F$, then $x\omega$ vanishes on $\mathcal{A}_F(D + \text{div}(x)) + F$, as mentioned before, and, therefore, $(\omega) + \text{div}(x) \leq (x\omega)$. Likewise, $(x\omega) + \text{div}(1/x) \leq (\omega)$. Combining these inequalities, we obtain $(\omega) + \text{div}(x) \leq (x\omega) \leq (\omega) - \text{div}(1/x) = (\omega) + \text{div}(x)$, which obviously yields the desired equality. $\square$

We are now ready for the climax of this section, the proof of the Riemann-Roch theorem. The first part of this theorem is a refinement of Riemann's theorem and the second part is a refinement of Corollary 3.5.3.

Theorem 3.6.14 (Riemann-Roch Theorem). Let W be a canonical divisor of the function field F/k with genus g. Then for any divisor $D \in \text{Div}(F)$ we have

$$\ell(D) = \deg(D) + 1 - g + \ell(W - D).$$

Furthermore, $\ell(D) = \deg(D) + 1 - g$ whenever $\deg(D) \geq 2g - 1$.

Proof. Let $W = (\omega)$ for some nonzero $\omega \in \Omega_F$ and consider the map $\lambda : \mathcal{L}(W - D) \to \Omega_F$ defined by $x \mapsto x\omega$. For nonzero $x \in \mathcal{L}(W - D)$, we have $(x\omega) = \text{div}(x) + (\omega) \geq -(W - D) + W = D$; hence, $x\omega \in \Omega_F(D)$. Therefore, λ maps $\mathcal{L}(W - D)$ into $\Omega_F(D)$. Clearly, λ is k-linear and injective. We claim that the image of λ is $\Omega_F(D)$. Consider a nonzero Weil differential $\omega_1 \in \Omega_F(D)$. Since Ω_F has dimension 1 over F by Theorem 3.6.10, there is $x \in F^*$ such that $\omega_1 = x\omega$. Now $\text{div}(x) + W = (x\omega) = (\omega_1) \geq D$; hence, $x \in \mathcal{L}(W - D)$ and $\omega_1 = \lambda(x)$, as desired. Thus, we have shown that $\mathcal{L}(W - D)$ and $\Omega_F(D)$ are isomorphic as vector spaces over k. By comparing dimensions and using Lemma 3.6.8, we obtain $\ell(W - D) = i(D)$, which settles the first part of the theorem by the definition of $i(D)$.

Putting $D = 0$, we get $\ell(W) = g$, and then putting $D = W$ yields $\deg(W) = 2g - 2$. For any $D \in \text{Div}(F)$ with $\deg(D) \geq 2g - 1$, we have $\deg(W - D) < 0$, and so $\ell(W - D) = 0$ by Corollary 3.4.4. This shows the second part of the theorem.   □

3.7 Elliptic Curves

In this section, we present a detailed treatment of nonsingular projective curves of genus 1 which are important, for instance, for applications to cryptography (see Section 6.2).

Definition 3.7.1. An *elliptic curve* over k is a pair $(\mathcal{E}, O)$, where $\mathcal{E}$ is a nonsingular projective curve over k of genus 1 and O is a k-rational point of $\mathcal{E}$. The point O corresponds to a rational place of the k-rational function field $k(\mathcal{E})$, which is called an *elliptic function field*.

For simplicity, we use the same symbol O to denote the rational place of $k(\mathcal{E})$ and the corresponding point $O \in \mathcal{E}$. The same applies to other k-rational points of $\mathcal{E}$. The correct interpretation will always be clear from the context.

Example 3.7.2.

(i) The affine plane curve over $\mathbb{F}_2$ defined by the equation

$$y^2 + y = x^3 + x$$

is smooth. We denote by O the point $[0, 1, 0]$ of the corresponding projective curve $\mathcal{E}$, that is, the curve defined in homogeneous coordinates by

$$y^2 z + y z^2 = x^3 + x z^2.$$

Then $v_O(x) = -2$ and $v_O(y) = -3$. For any integer $m \geq 0$, the set

$$S_m := \{x^i y^j : i \geq 0, \ 0 \leq j \leq 1, \ 2i + 3j \leq 2m + 3\}$$

forms an $\mathbb{F}_2$-basis of the Riemann-Roch space $\mathcal{L}((2m + 3)O)$. The set S_m has exactly $2m + 3$ elements. Hence, $\ell((2m + 3)O) = |S_m| = 2m + 3$. By Corollary 3.5.3, the genus of $\mathcal{E}$ is 1, that is, the pair $(\mathcal{E}, O)$ is an elliptic curve. There are altogether five $\mathbb{F}_2$-rational points of $\mathcal{E}$, namely

$$[0, 0, 1], \ [0, 1, 1], \ [1, 0, 1], \ [1, 1, 1], \ \text{and} \ O.$$

(ii) The affine plane curve over $\mathbb{F}_3$ defined by the equation

$$y^2 = x^3 - x + 1$$

is smooth. We denote by O the point $[0, 1, 0]$ of the corresponding projective curve $\mathcal{E}$. As in (i), we see that the pair $(\mathcal{E}, O)$ is an elliptic curve. There are altogether seven $\mathbb{F}_3$-rational points of $\mathcal{E}$, namely

$$[0, 1, 1], \ [0, 2, 1], \ [1, 1, 1], \ [1, 2, 1], \ [2, 1, 1], \ [2, 2, 1], \ \text{and} \ O.$$

Theorem 3.7.3. Let $(\mathcal{E}, O)$ be an elliptic curve over k and let $\mathcal{E}(k)$ denote the set of k-rational points of $\mathcal{E}$. Let $F = k(\mathcal{E})$ be the k-rational function field of $\mathcal{E}$. Then the map

$$\chi : \mathcal{E}(k) \rightarrow \mathrm{Div}^0(F)/\mathrm{Princ}(F), \quad P \mapsto \overline{P - O},$$

is a bijection. Here $\overline{D}$ denotes the image of $D \in \mathrm{Div}^0(F)$ under the canonical homomorphism from $\mathrm{Div}^0(F)$ to $\mathrm{Div}^0(F)/\mathrm{Princ}(F)$.

Proof. Let P and Q be two distinct k-rational points of $\mathcal{E}$ with $\chi(P) = \chi(Q)$, that is, for the corresponding places O, P, and Q of F we have $P - O \sim Q - O$. This implies $P \sim Q$, and so

$\mathrm{div}(w) = P - Q$ for some $w \in F^*$. Then $(w)_0 = P$ and $\deg((w)_0) = 1$; hence, $F = k(w)$ by Proposition 3.4.5. But then F has genus 0 by Example 3.5.5, which is a contradiction. This shows that χ is injective.

Now for any divisor $D \in \mathrm{Div}^0(F)$, the Riemann-Roch space $\mathcal{L}(D + O)$ has dimension 1 by the Riemann-Roch theorem. Let z be a nonzero element of $\mathcal{L}(D + O)$. Then $\mathrm{div}(z) + D + O$ is a positive divisor of degree 1. Hence, it must be a rational place of F and corresponds to a point $P \in \mathcal{E}(k)$ for which we get $\chi(P) = \overline{D}$. Thus, χ is surjective. $\square$

Since χ in Theorem 3.7.3 is bijective, we can equip the set $\mathcal{E}(k)$ of k-rational points of $\mathcal{E}/k$ with the structure of an abelian group by defining the addition $\oplus$ in $\mathcal{E}(k)$ through χ, that is,

$$P \oplus Q = \chi^{-1}(\overline{P + Q - 2O}) \in \mathcal{E}(k)$$

for any two k-rational points $P, Q \in \mathcal{E}(k)$. The groups $\mathcal{E}(k)$ and $\mathrm{Div}^0(F)/\mathrm{Princ}(F)$ are then isomorphic and O is the identity element of $\mathcal{E}(k)$.

For any integer $m \geq 1$, we denote by $[m]P$ the point in $\mathcal{E}(k)$ given by the sum

$$[m]P = \underbrace{P \oplus \cdots \oplus P}_{m}.$$

It is convenient to denote the inverse of $[m]P$ in the group $\mathcal{E}(k)$ by $[-m]P$. Furthermore, we set $[0]P = O$. By Theorem 3.7.3, we have the following immediate result.

Corollary 3.7.4. Let $(\mathcal{E}, O)$ be an elliptic curve over k with k-rational function field F and let $D = \sum_{P \in \mathbf{P}_F} n_P P$ be a divisor of F. Then D is principal if and only if $\deg(D) = 0$ and $\oplus \sum [n_P]P = O$ (note that we consider the addition $\oplus$ of $\mathcal{E}(K)$ for some extension K/k such that $\mathrm{supp}(D) \subseteq \mathcal{E}(K)$ in an obvious interpretation).

Example 3.7.5. In Example 3.7.2(i), we have

$$[0, 0, 1] \oplus [0, 1, 1] = O$$

as $\mathrm{div}(x) = [0, 0, 1] + [0, 1, 1] - 2O$. In Example 3.7.2(ii), we have

$$[0, 1, 1] \oplus [1, 1, 1] \oplus [2, 1, 1] = O$$

as $\mathrm{div}(y) = [0, 1, 1] + [1, 1, 1] + [2, 1, 1] - 3O$.

Let $\mathcal{X}/k$ be an affine plane curve defined by a *Weierstrass equation*

$$y^2 + a_1 xy + a_3 y = x^3 + a_2 x^2 + a_4 x + a_6 \tag{3.8}$$

for some $a_1, a_2, a_3, a_4, a_6 \in k$. Define the *discriminant* of $\mathcal{X}$ by

$$\Delta(\mathcal{X}) = -b_2^2 b_8 - 8 b_4^3 - 27 b_6^2 + 9 b_2 b_4 b_6, \tag{3.9}$$

where

$$b_2 = a_1^2 + 4a_2, \quad b_4 = 2a_4 + a_1 a_3, \quad b_6 = a_3^2 + 4a_6,$$
$$b_8 = a_1^2 a_6 + 4a_2 a_6 - a_1 a_3 a_4 + a_2 a_3^2 - a_4^2. \tag{3.10}$$

Proposition 3.7.6. Let $\mathcal{X}/k$ be the affine plane curve defined by the Weierstrass equation (3.8), let $\mathcal{E}$ be the corresponding projective curve, and let O be the point $[0, 1, 0]$ of $\mathcal{E}$. Then:

 (i) $\mathcal{E}$ is smooth if and only if $\Delta(\mathcal{X}) \neq 0$;
 (ii) if $\mathcal{E}$ is smooth, then $(\mathcal{E}, O)$ is an elliptic curve;
 (iii) if $\mathcal{E}$ is not smooth, then $\mathcal{E}$ is birationally equivalent to the projective line, that is, the function field of $\mathcal{E}$ is isomorphic to a rational function field.

Proof.

 (i) If $h(x, y, z) = 0$ is (3.8) in homogeneous coordinates, then it is easy to verify that $\frac{\partial h}{\partial z}\big|_{(0,1,0)} \neq 0$, and so O is a smooth point. To simplify the computation, we may assume that the characteristic of k is not 2. Replacing y in (3.8) by $(y - a_1 x - a_3)/2$, we obtain the equation

$$y^2 = 4x^3 + b_2 x^2 + 2b_4 x + b_6, \tag{3.11}$$

where b_2, b_4, b_6 are defined in (3.10). Thus, by Example 3.1.4 a point $P = (x_0, y_0)$ of $\mathcal{X}$ is singular if and only if

$$2y_0 = 0 = 12x_0^2 + 2b_2x_0 + 2b_4.$$

If we put $f(x) = 4x^3 + b_2x^2 + 2b_4x + b_6$, then P is singular if and only if x_0 satisfies $f(x_0) = 0$ and $f'(x_0) = 0$, that is, x_0 is a double root of $f(x)$. This cubic polynomial has a double root if and only if its discriminant $16\Delta(\mathcal{X})$ is zero.

(ii) By checking the dimension of $\mathcal{L}(mO)$ as in Example 3.7.2, we see that the genus of $\mathcal{E}$ is 1.

(iii) Again for simplicity we assume that the characteristic of k is not 2 and that the curve is defined by the equation (3.11). Since $f(x) = 4x^3 + b_2x^2 + 2b_4x + b_6$ has a double root x_0, we may write $f(x) = (x - x_0)^2(4x + b)$. Hence, (3.11) becomes $(y/(x - x_0))^2 = 4x + b$ and the function field of $\mathcal{E}$ is $k(x, y) = k(x, y/(x - x_0)) = k(t)$ with $t = y/(x - x_0)$. By Corollary 2.5.25, $\mathcal{E}$ is birationally equivalent to the projective line.   □

Theorem 3.7.7. Let $(\mathcal{E}, O)$ be an elliptic curve over k. Then there exists an isomorphism ϕ from $\mathcal{E}/k$ to a projective plane curve $\mathcal{Y}/k$ defined by the homogenized form of a Weierstrass equation

$$y^2 + a_1xy + a_3y = x^3 + a_2x^2 + a_4x + a_6$$

with $a_1, a_2, a_3, a_4, a_6 \in k$ and $\phi(O) = [0, 1, 0]$.

Proof. By the Riemann-Roch theorem, we have $\ell(iO) = i$ for all integers $i \geq 1$. Hence, there exist two elements x and y of $k(\mathcal{E})$ such that $(x)_\infty = 2O$ and $(y)_\infty = 3O$. The seven elements $1, x, y, x^2, xy, x^3, y^2$ all belong to the Riemann-Roch space $\mathcal{L}(6O)$ of dimension 6; hence, there is a linear relation

$$s_0 + s_1x + s_2y + s_3x^2 + s_4xy + s_5x^3 + s_6y^2 = 0 \qquad (3.12)$$

for some $s_i \in k$ that are not all 0. Note that both s_5 and s_6 are nonzero, since otherwise we get a contradiction by considering the values of v_O at the terms in (3.12). Replacing x and y by $-s_5s_6x$ and $s_5^2s_6y$, respectively, and dividing by $s_5^4s_6^3$ in (3.12) yields the desired cubic equation in Weierstrass form.

Now we prove that $k(\mathcal{E}) = k(x, y)$. By Theorem 3.4.2 and Corollary 3.4.3, we have $[k(\mathcal{E}) : k(x)] = 2$ and $[k(\mathcal{E}) : k(y)] = 3$ since $\deg((x)_0) = \deg((x)_\infty) = 2$ and $\deg((y)_0) = \deg((y)_\infty) = 3$. Therefore, $[k(\mathcal{E}) : k(x, y)] = 1$ since 2 and 3 are coprime.

From the above we get a rational map $\phi : \mathcal{E} \to \mathcal{Y}$ of degree 1 with $\phi(O) = [0, 1, 0]$. Note that ϕ is a morphism by Proposition 3.2.6.

Next we show that $\mathcal{Y}$ is smooth. Suppose $\mathcal{Y}$ were not smooth. By Proposition 3.7.6(iii), there exists a rational map $\varphi : \mathcal{Y} \to \mathbf{P}^1$ of degree 1. Hence, the composition $\varphi \circ \phi : \mathcal{E} \to \mathbf{P}^1$ is a rational map of degree 1. As $\mathcal{E}$ and $\mathbf{P}^1$ are both smooth curves, this composition is an isomorphism by [114, Corollary II.2.4.1]. This contradicts the fact that two isomorphic curves have the same genus as the genus of $\mathcal{E}$ is 1 and the genus of $\mathbf{P}^1$ is 0. Thus, $\mathcal{Y}$ is indeed smooth, and so another application of [114, Corollary II.2.4.1] shows that ϕ is an isomorphism.   □

In other words, Theorem 3.7.7 says that any elliptic curve can be expressed by a Weierstrass equation. From now on we will focus on Weierstrass equations, where it will always be understood that the elliptic curve is actually defined by the homogenized form of the Weierstrass equation. We now consider the group law of an elliptic curve defined by a Weierstrass equation.

Proposition 3.7.8. Let $(\mathcal{E}, O)$ be an elliptic curve over k defined by the Weierstrass equation

$$y^2 + a_1 xy + a_3 y = x^3 + a_2 x^2 + a_4 x + a_6$$

with $O = [0, 1, 0]$. Then three points $P_1, P_2, P_3 \in \mathcal{E}(k)$ (we do not require that P_1, P_2, P_3 be distinct) satisfy $P_1 \oplus P_2 \oplus P_3 = O$ if and only if

$$\operatorname{div}(ay + bx + c) = P_1 + P_2 + P_3 - 3O$$

for some $a, b, c \in k$ that are not all zero.

Proof. By Corollary 3.7.4, $P_1 \oplus P_2 \oplus P_3 = O$ if and only if $P_1 + P_2 + P_3 - 3O = \operatorname{div}(z)$ for some $z \neq 0$. As z belongs to $\mathcal{L}(3O)$ and $\{1, x, y\}$ is a k-basis of $\mathcal{L}(3O)$, the desired result follows.   □

The above result can be converted to get the following explicit algebraic formula for the sum of two points.

Corollary 3.7.9. Let $(\mathcal{E}, O)$ be an elliptic curve over k defined by the Weierstrass equation

$$y^2 + a_1 x y + a_3 y = x^3 + a_2 x^2 + a_4 x + a_6$$

with $O = [0, 1, 0]$. Let $P_1 = (x_1, y_1)$ and $P_2 = (x_2, y_2)$ be two points of $\mathcal{E}(k)$ and let $P_1 \oplus P_2 = P_0$.

(i) If $P_0 = O$, then

$$(x_2, \ y_2) = (x_1, \ -y_1 - a_1 x_1 - a_3).$$

(ii) If $P_0 = (x_0, y_0) \neq O$, then

$$(x_0, \ y_0) = (\lambda^2 + a_1 \lambda - a_2 - x_1 - x_2, \ -(\lambda + a_1)x_0 - \mu - a_3),$$

where

$$\lambda = \begin{cases} \dfrac{y_2 - y_1}{x_2 - x_1} & \text{if } x_1 \neq x_2, \\[2ex] \dfrac{3x_1^2 + 2a_2 x_1 + a_4 - a_1 y_1}{2y_1 + a_1 x_1 + a_3} & \text{if } x_1 = x_2, \end{cases}$$

and

$$\mu = \begin{cases} \dfrac{y_1 x_2 - y_2 x_1}{x_2 - x_1} & \text{if } x_1 \neq x_2, \\[2ex] \dfrac{-x_1^3 + a_4 x_1 + 2a_6 - a_3 y_1}{2y_1 + a_1 x_1 + a_3} & \text{if } x_1 = x_2. \end{cases}$$

Proposition 3.7.10. Two Weierstrass equations defining the same elliptic curve $(\mathcal{E}, O)$ with O corresponding to $[0, 1, 0]$ are related by the substitution of variables

$$\begin{cases} x = u^2 x' + r \\ y = u^3 y' + s u^2 x' + t \end{cases}$$

for some $u, r, s, t \in k$ with $u \neq 0$.

Proof. Let $\{x, y\}$ and $\{x', y'\}$ be two sets of Weierstrass coordinate functions on $\mathcal{E}$. Then $v_O(x) = v_O(x') = -2$ and $v_O(y) = v_O(y') = -3$. Hence, both $\{1, x\}$ and $\{1, x'\}$ are k-bases for $\mathcal{L}(2O)$, and similarly both $\{1, x, y\}$ and $\{1, x', y'\}$ are k-bases for $\mathcal{L}(3O)$. It follows that there are elements $u_1, u_2, r, s_2, t \in k$ with $u_1 u_2 \neq 0$ such that

$$x = u_1 x' + r, \qquad y = u_2 y' + s_2 x' + t.$$

Comparing coefficients after replacing x and y according to the above relations, we obtain $u_1^3 = u_2^2$. Put $u = u_2/u_1$ and $s = s_2/u^2$, and then we get the desired substitution. $\qquad\square$

Using a suitable linear substitution, we can simplify the form of the Weierstrass equation for an elliptic curve.

Theorem 3.7.11. Under a substitution

$$\begin{cases} x = u^2 x' + r \\ y = u^3 y' + su^2 x' + t \end{cases}$$

for some $u, r, s, t \in k$ with $u \neq 0$, a Weierstrass equation

$$y^2 + a_1 xy + a_3 y = x^3 + a_2 x^2 + a_4 x + a_6$$

over k can be simplified to the following form:

(i) if char$(k) \neq 2, 3$, then

$$y^2 = x^3 + Ax + B$$

for some $A, B \in k$ and, furthermore, it defines an elliptic curve if and only if $\triangle := -16(4A^3 + 27B^2) \neq 0$;

(ii) if char$(k) = 2$, then

$$y^2 + xy = x^3 + Ax^2 + B$$

for some $A, B \in k$ and, furthermore, it defines an elliptic curve if and only if $\triangle := B \neq 0$; or

$$y^2 + Cy = x^3 + Ax^2 + B$$

for some $A, B \in k$ and, furthermore, it defines an elliptic curve if and only if $\triangle := C^4 \neq 0$;

(iii) if char$(k) = 3$, then

$$y^2 = x^3 + Ax^2 + B$$

for some $A, B \in k$ and, furthermore, it defines an elliptic curve if and only if $\triangle := -A^3 B \neq 0$; or

$$y^2 = x^3 + Ax + B$$

for some $A, B \in k$ and, furthermore, it defines an elliptic curve if and only if $\triangle := -A^3 \neq 0$.

Proof.

(i) If char$(k) \neq 2$, then we have already seen in the proof of Proposition 3.7.6(i) that a Weierstrass equation can be put in the form (3.11). Furthermore, if also char$(k) \neq 3$, then replacing x by $x + c$ and y by $2y$ for a suitable $c \in k$ yields a cubic equation of the desired form.

(ii) Let char$(k) = 2$ and recall that k is perfect, so that square roots exist in k. If $a_1 = 0$, then replacing x by $x + c$ for a suitable $c \in k$ yields a cubic equation of the desired form $y^2 + Cy = x^3 + Ax^2 + B$ with $C = a_3$ and some $A, B \in k$. If $a_1 \neq 0$, then replacing x by $ax + c$ and y by $by + dx + e$ for suitable $a, b, c, d, e \in k$ yields a cubic equation of the desired form $y^2 + xy = x^3 + Ax^2 + B$ for some $A, B \in k$.

(iii) If char$(k) = 3$, then starting from (3.11) and replacing x by $x + c$ for a suitable $c \in k$ yields a cubic equation with either the coefficient of x or the coefficient of x^2 being zero. In all cases (i), (ii), and (iii), the criterion for obtaining an elliptic curve, that is, the criterion for smoothness, follows from Proposition 3.7.6(i) and the formula (3.9). $\square$

There are several books devoted specifically to elliptic curves, some of them treating also applications to cryptography. We mention Blake,

TABLE 3.8.1
Dictionary Curve—Function Field

Curve	Function field
$\mathbb{F}_q$-closed point	Place
$\mathbb{F}_q$-rational point	Rational place
Degree (size) of $\mathbb{F}_q$-closed point	Degree of place
Genus of curve	Genus of function field
Projective line	Rational function field
Covering	Finite extension

Seroussi, and Smart [9], Enge [32], Husemoller [58], Silverman [114], and Washington [126].

3.8 Summary: Curves and Function Fields

One of the salient features of this chapter is the interplay between curves and function fields, which allows us to move with ease between these mathematical objects. For the convenience of the reader, we summarize these relationships in a compact form in Table 3.8.1 for the case where k is a finite field. This case is of great relevance for the later chapters.

For all entries in Table 3.8.1, except for the last one, a rationale was given in earlier parts of this chapter. Note that in Table 3.8.1 "curve" stands for nonsingular projective curve over $\mathbb{F}_q$ and "function field" stands for global function field with full constant field $\mathbb{F}_q$.

To explain the last entry, we recall from Theorem 3.2.3 that certain nonconstant morphisms between nonsingular projective curves over $\mathbb{F}_q$ induce homomorphisms between their $\mathbb{F}_q$-rational function fields. This procedure can be reversed (see [114, Section II.2]). Suppose we are given two global function fields $F_1/\mathbb{F}_q$ and $F_2/\mathbb{F}_q$ with $F_2 \subseteq F_1$. Note that $[F_1 : F_2] < \infty$. Then from the embedding of F_2 into F_1, we obtain a surjective morphism $\phi : \mathcal{X}_1/\mathbb{F}_q \to \mathcal{X}_2/\mathbb{F}_q$ with the property that the fiber $\phi^{-1}(Q)$ has cardinality $|\phi^{-1}(Q)| \le [F_1 : F_2]$ for every point $Q \in \mathcal{X}_2$. We speak in this case of a *covering* $\mathcal{X}_1 \to \mathcal{X}_2$. Concepts referring to field extensions can be transferred to coverings. For instance, $\mathcal{X}_1 \to \mathcal{X}_2$ is a *Galois covering* if F_1/F_2 is a Galois extension. If $|\phi^{-1}(Q)| = [F_1 : F_2]$, then a Galois covering $\mathcal{X}_1 \to \mathcal{X}_2$ is called *unramified* at $Q \in \mathcal{X}_2$, otherwise it is called *ramified* at Q. A Galois covering $\mathcal{X}_1 \to \mathcal{X}_2$ can be ramified at only finitely many points of $\mathcal{X}_2$.

4   Rational Places

Many applications of algebraic curves over a finite field $\mathbb{F}_q$ utilize the $\mathbb{F}_q$-rational points of these curves. Thus, it is of great interest to study these points. Since we are adopting the equivalent viewpoint of global function fields, we consider instead rational places of global function fields, that is, places of degree 1. A major result on rational places is the Hasse-Weil bound on the number of rational places of a global function field. This bound is derived in Section 4.2 from the equally important Hasse-Weil theorem on zeta functions of global function fields. Preliminaries on zeta functions and divisor class numbers of global function fields are presented in Section 4.1. Refinements of the Hasse-Weil bound and asymptotic results on the number of rational places are discussed in Section 4.3. Applications of the Hasse-Weil bound to the estimation of character sums over finite fields are shown in Section 4.4.

4.1 Zeta Functions

Let $F/\mathbb{F}_q$ be a global function field with full constant field $\mathbb{F}_q$. We recall the notation $\mathbf{P}_F$ for the set of places of F and $\mathrm{Div}(F)$ for the divisor group of F.

Definition 4.1.1. For any integer $n \geq 0$, let $A_n(F)$ be the cardinality of the set of divisors $D \in \mathrm{Div}(F)$ with $D \geq 0$ and $\deg(D) = n$.

Proposition 4.1.2. $A_n(F)$ is finite for any $n \geq 0$.

Proof. From the representation $D = \sum_{P \in \mathbf{P}_F} n_P P$, $n_P \geq 0$, for a positive divisor $D \in \mathrm{Div}(F)$ and the definition of $\deg(D)$, we see that it suffices to show that there exist only finitely many places of F of a given degree d. Due to the connections between function fields and curves (see Section 3.8), a place of $F/\mathbb{F}_q$ of degree d corresponds to an $\mathbb{F}_q$-closed point P of degree d of a suitable nonsingular projective curve $\mathcal{X}/\mathbb{F}_q$. By Lemma 2.1.11(iv), the definition field $\mathbb{F}_q(\mathsf{P})$ of P is given by $\mathbb{F}_{q^d}$. Therefore, $\mathsf{P} \subseteq \mathbf{P}^m(\mathbb{F}_{q^d})$ for some $m \geq 1$ depending only on $\mathcal{X}$, and so Proposition 2.1.8 shows that there are only finitely many choices for P. $\qquad\qquad\square$

We remind the reader that a place of $F/\mathbb{F}_q$ of degree 1 is also called a rational place of $F/\mathbb{F}_q$ and that the rational places of $F/\mathbb{F}_q$ are in one-to-one correspondence with the $\mathbb{F}_q$-rational points of the corresponding nonsingular projective curve over $\mathbb{F}_q$. The following result is an immediate consequence of Proposition 4.1.2.

Corollary 4.1.3. A global function field has only finitely many rational places.

Let $N(F)$ denote the number of rational places of $F/\mathbb{F}_q$. More generally, for each integer $n \geq 1$ we consider the constant field extension $F_n := F \cdot \mathbb{F}_{q^n}$ and let $N_n(F)$ be the number of rational places of $F_n/\mathbb{F}_{q^n}$. Thus, we have $N(F) = N_1(F)$. The zeta function of $F/\mathbb{F}_q$ is an important device for incorporating all these data.

Definition 4.1.4. The *zeta function* $Z(F, t)$ of $F/\mathbb{F}_q$ is the formal power series

$$Z(F, t) := \exp\left(\sum_{n=1}^{\infty} \frac{N_n(F)}{n} t^n\right) \in \mathbb{C}[[t]]$$

over the complex numbers.

Example 4.1.5. We compute the zeta function of the rational function field $F = \mathbb{F}_q(x)$ over $\mathbb{F}_q$. For all $n \geq 1$, we have $F_n = \mathbb{F}_{q^n}(x)$, and so $N_n(F) = q^n + 1$ by Example 1.5.15. Hence, we get

$$\log Z(F, t) = \sum_{n=1}^{\infty} \frac{q^n + 1}{n} t^n = \sum_{n=1}^{\infty} \frac{(qt)^n}{n} + \sum_{n=1}^{\infty} \frac{t^n}{n}$$

$$= -\log(1 - qt) - \log(1 - t) = \log \frac{1}{(1 - t)(1 - qt)},$$

that is,

$$Z(F, t) = \frac{1}{(1 - t)(1 - qt)}.$$

Theorem 4.1.6. The zeta function $Z(F, t)$ can be represented also by the following two expressions:

(i) $Z(F, t) = \prod_{P \in \mathbb{P}_F} (1 - t^{\deg(P)})^{-1}$;
(ii) $Z(F, t) = \sum_{n=0}^{\infty} A_n(F) t^n$.

Proof.

(i) In view of Proposition 4.1.2, the cardinality $B_d(F)$ of the set of places of F of fixed degree $d \geq 1$ is finite. By Theorem 1.7.2, a place of F of degree d splits into rational places of F_n if and only if d divides n. In the case where d divides n, a place of F of degree d splits into exactly d rational places of F_n. Therefore, we get the identity

$$N_n(F) = \sum_{d \mid n} d \, B_d(F). \tag{4.1}$$

Now we have

$$\log \left(\prod_{P \in \mathbb{P}_F} (1 - t^{\deg(P)})^{-1} \right) = \log \left(\prod_{d=1}^{\infty} (1 - t^d)^{-B_d(F)} \right)$$

$$= -\sum_{d=1}^{\infty} B_d(F) \log(1 - t^d)$$

$$= \sum_{d=1}^{\infty} B_d(F) \sum_{n=1}^{\infty} \frac{t^{dn}}{n}$$

$$= \sum_{n=1}^{\infty} \sum_{d \mid n} \frac{d \, B_d(F)}{n} t^n = \sum_{n=1}^{\infty} \frac{N_n(F)}{n} t^n,$$

where we used (4.1) in the last step. Therefore,

$$Z(F, t) = \exp \left(\sum_{n=1}^{\infty} \frac{N_n(F)}{n} t^n \right) = \prod_{P \in \mathbb{P}_F} (1 - t^{\deg(P)})^{-1}.$$

(ii) From (i) we get

$$Z(F, t) = \prod_{P \in \mathbf{P}_F} \left(1 - t^{\deg(P)}\right)^{-1} = \prod_{P \in \mathbf{P}_F} \left(\sum_{n=0}^{\infty} t^{\deg(nP)}\right)$$

$$= \sum_{D \in \mathrm{Div}(F),\ D \geq 0} t^{\deg(D)} = \sum_{n=0}^{\infty} A_n(F) t^n,$$

by collecting terms appropriately in the last step. Note that the last two formal power series make sense because of Proposition 4.1.2. $\qquad \square$

The constant field extension $F_n = F \cdot \mathbb{F}_{q^n}$ is again a global function field, and so we can consider its zeta function $Z(F_n, t)$. There is a close relationship between $Z(F_n, t)$ and $Z(F, t)$, which is enunciated in the following result.

Proposition 4.1.7. For every positive integer n we have

$$Z(F_n, t^n) = \prod_{\zeta^n = 1} Z(F, \zeta t),$$

where the product is extended over all complex nth roots of unity ζ.

Proof. By Theorem 4.1.6(i), we have

$$Z(F_n, t^n) = \prod_{Q \in \mathbf{P}_{F_n}} \left(1 - t^{n \deg(Q)}\right)^{-1} = \prod_{P \in \mathbf{P}_F} \prod_{Q \mid P} \left(1 - t^{n \deg(Q)}\right)^{-1}.$$

For a fixed place $P \in \mathbf{P}_F$ of degree d, we know by Theorem 1.7.2 that the degree of the places Q of F_n lying over P is $d/\gcd(d, n)$ and that there are exactly $\gcd(d, n)$ places of F_n lying over P. Hence,

$$\prod_{Q \mid P} \left(1 - t^{n \deg(Q)}\right) = \left(1 - (t^d)^{n/\gcd(d,n)}\right)^{\gcd(d,n)}$$

$$= \prod_{\zeta^n = 1} \left(1 - \zeta^d t^d\right) = \prod_{\zeta^n = 1} \left(1 - (\zeta t)^{\deg(P)}\right).$$

This yields

$$Z(F_n, t^n) = \prod_{\zeta^n = 1} \prod_{P \in \mathbf{P}_F} \left(1 - (\zeta t)^{\deg(P)}\right)^{-1} = \prod_{\zeta^n = 1} Z(F, \zeta t)$$

by Theorem 4.1.6(i). $\qquad \square$

We recall from Section 3.4 that two divisors D and G of F are said to be equivalent if $G = D + \text{div}(x)$ for some $x \in F^*$. The *divisor class* $[D] := D + \text{Princ}(F)$ consists of all divisors of F that are equivalent to D. Note that by Corollary 3.4.3, all divisors in the divisor class $[D]$ have the same degree $\deg(D)$, which we can therefore call the *degree* of the divisor class. Addition and subtraction of divisor classes are well defined since the divisor classes form the factor group $\text{Div}(F)/\text{Princ}(F)$.

The subset $\text{Div}^0(F)$ of $\text{Div}(F)$ consisting of all divisors of F of degree 0 is a subgroup of $\text{Div}(F)$, and in turn $\text{Princ}(F)$ is a subgroup of $\text{Div}^0(F)$. Thus, we can consider the factor group

$$\text{Cl}(F) := \text{Div}^0(F)/\text{Princ}(F),$$

which is called the *group of divisor classes of degree 0* of F. In other words, $\text{Cl}(F)$ consists of all divisor classes $[D]$ with $D \in \text{Div}^0(F)$.

Proposition 4.1.8. $\text{Cl}(F)$ is a finite abelian group.

Proof. We need to show only that $\text{Cl}(F)$ is finite. Choose $D \in \text{Div}(F)$ with $d := \deg(D) \geq g$, where g is the genus of F. By Riemann's theorem (see Theorem 3.5.1), we have $\ell(D) \geq d + 1 - g \geq 1$, and so the definition of $\ell(D)$ shows that there exists a positive divisor G of F with $[G] = [D]$. But since $A_d(F)$ is finite by Proposition 4.1.2, there are only finitely many divisor classes of degree d of F. The divisor classes of degree 0 of F are given exactly by all $[B] - [D]$ with $B \in \text{Div}(F)$ of degree d, and so $\text{Cl}(F)$ is finite. $\qquad\square$

The order of the finite group $\text{Cl}(F)$ is denoted by $h(F)$ and called the *divisor class number* of F. Note that for a rational function field every divisor of degree 0 is principal, and so the divisor class number of a rational function field is equal to 1.

We are now going to prove that the zeta function $Z(F, t)$ of a global function field $F/\mathbb{F}_q$ is, in fact, a rational function in t of a special form. First we need the following auxiliary results.

Lemma 4.1.9. For any $D \subset \text{Div}(F)$, the number of positive divisors in the divisor class $[D]$ is given by

$$\frac{q^{\ell(D)} - 1}{q - 1}.$$

Proof. A divisor $G \in [D]$ is positive if and only if $G = D + \operatorname{div}(x)$ for some $x \in \mathcal{L}(D)\backslash\{0\}$. There are exactly $q^{\ell(D)} - 1$ elements $x \in \mathcal{L}(D)\backslash\{0\}$, and $x, y \in \mathcal{L}(D)\backslash\{0\}$ yield the same divisor G if and only if $y = cx$ for some $c \in \mathbb{F}_q^*$. □

Lemma 4.1.10. Any global function field has a divisor of degree 1. In particular, any global function field of genus 0 is a rational function field.

Proof. Given a global function field $F/\mathbb{F}_q$ of genus g, let the image of the group homomorphism $D \in \operatorname{Div}(F) \mapsto \deg(D)$ be the subgroup of $\mathbb{Z}$ generated by $k \geq 1$. For the first part of the lemma, we have to show that $k = 1$.

For every integer $n \geq 0$ with $k|n$, we have exactly $h(F)$ divisor classes $[D_1], [D_2], \ldots, [D_{h(F)}]$ of degree n. Thus by Lemma 4.1.9,

$$A_n(F) = \sum_{i=1}^{h(F)} \frac{q^{\ell(D_i)} - 1}{q - 1}.$$

Therefore, if $n \geq \max(0, 2g - 1)$ and $k|n$, then

$$A_n(F) = \frac{h(F)}{q - 1}(q^{n+1-g} - 1) \tag{4.2}$$

by the Riemann-Roch theorem (see Theorem 3.6.14). Together with Theorem 4.1.6(ii) this shows that

$$(q - 1)\, Z(F, t) = f_1(t^k) + h(F) \sum_{\substack{n=2g \\ k|n}}^{\infty} (q^{n+1-g} - 1)\, t^n$$

$$= f_2(t^k) + h(F) \sum_{\substack{n=0 \\ k|n}}^{\infty} (q^{n+1-g} - 1)\, t^n$$

$$= f_2(t^k) + h(F) q^{1-g} \sum_{n=0}^{\infty} (q^k t^k)^n - h(F) \sum_{n=0}^{\infty} (t^k)^n$$

$$= f_2(t^k) + \frac{h(F)\, q^{1-g}}{1 - q^k\, t^k} - \frac{h(F)}{1 - t^k}$$

with polynomials f_1 and f_2. Thus, $Z(F, t)$ is a rational function with a simple pole at $t = 1$, and for the same reason $Z(F_k, t^k)$ also has a simple pole at $t = 1$. Furthermore, for any complex kth root of unity ζ we have $Z(F, \zeta t) = Z(F, t)$; hence,

$$Z(F_k, t^k) = Z(F, t)^k$$

by Proposition 4.1.7. Consequently, $Z(F_k, t^k)$ has a pole of order k at $t = 1$. From the earlier statement about the pole of $Z(F_k, t^k)$ at $t = 1$ it follows now that $k = 1$.

Now let $F/\mathbb{F}_q$ have genus 0. By what we have just shown, there exists a divisor D of F with $\deg(D) = 1$. Choose a positive divisor $G \in [D]$, which exists by Lemma 4.1.9 since $\ell(D) = 2$ by the Riemann-Roch theorem. Then $G \geq 0$ and $\deg(G) = 1$; hence, G is a rational place of F with $\ell(G) = 2$, and so Proposition 3.4.5 shows that F is a rational function field. □

Theorem 4.1.11. Let $F/\mathbb{F}_q$ be a global function field of genus g. Then the zeta function $Z(F, t)$ of F is a rational function of the form

$$Z(F, t) = \frac{L(F, t)}{(1 - t)(1 - qt)},$$

where $L(F, t) \in \mathbb{Z}[t]$ is a polynomial of degree at most $2g$ with integral coefficients and $L(F, 0) = 1$. Moreover, $L(F, 1)$ is equal to the divisor class number $h(F)$ of F.

Proof. If $g = 0$, then F is a rational function field by Lemma 4.1.10, and so the result follows from Example 4.1.5. If $g \geq 1$, then we can use (4.2) for all $n \geq 2g - 1$ (note that $k = 1$ by Lemma 4.1.10). Thus, a calculation similar to that following (4.2) yields

$$(q - 1)Z(F, t) = f(t) + \frac{h(F)q^{1-g}}{1 - qt} - \frac{h(F)}{1 - t} \qquad (4.3)$$

with a polynomial f satisfying $\deg(f) \leq 2g - 2$. Thus,

$$Z(F, t) = \frac{L(F, t)}{(1 - t)(1 - qt)}$$

with a polynomial $L(F, t)$ of degree at most $2g$. A comparison with Theorem 4.1.6(ii) shows that $L(F, t) \in \mathbb{Z}[t]$. Since $Z(F, 0) = 1$ by Definition 4.1.4, we get $L(F, 0) = 1$. Finally, a comparison with (4.3) yields $L(F, 1) = h(F)$. □

Definition 4.1.12. The polynomial $L(F, t) = (1 - t)(1 - qt)Z(F, t)$ is called the *L-polynomial* of $F/\mathbb{F}_q$.

The *L*-polynomial $L(F, t)$ has further important properties apart from those listed in Theorem 4.1.11. Some of them can be deduced from its functional equation.

Theorem 4.1.13. The *L*-polynomial $L(F, t)$ of the global function field $F/\mathbb{F}_q$ of genus g satisfies the functional equation

$$L(F, t) = q^g t^{2g} L\left(F, \frac{1}{qt}\right).$$

Proof. Since the case $g = 0$ is trivial, we can assume that $g \geq 1$. It suffices to show that the function $t^{1-g} Z(F, t)$ is invariant under the transformation $t \mapsto q^{-1}t^{-1}$. By Lemma 4.1.9, we can write

$$A_n(F) = \sum_{\deg([D])=n} \frac{q^{\ell(D)} - 1}{q - 1} \quad \text{for all } n \geq 0.$$

Since there are $h(F)$ divisor classes of any fixed degree n, we obtain with $\mathcal{D} := \{[D] : 0 \leq \deg([D]) \leq 2g - 2\}$,

$$(q - 1)t^{1-g}Z(F, t) = t^{1-g} \sum_{n=0}^{\infty} \left(\sum_{\deg([D])=n} (q^{\ell(D)} - 1) \right) t^n$$

$$= \sum_{[D]\in\mathcal{D}} q^{\ell(D)} t^{\deg([D])+1-g} - \frac{h(F)t^{1-g}}{1-t}$$

$$+ h(F) \sum_{n=2g-1}^{\infty} q^{n+1-g} t^{n+1-g}$$

$$= X(t) + Y(t),$$

where

$$X(t) = \sum_{[D]\in\mathcal{D}} q^{\ell(D)} t^{\deg([D])+1-g},$$

$$Y(t) = \frac{h(F)q^g t^g}{1 - qt} - \frac{h(F)t^{1-g}}{1 - t}.$$

It is immediate that $Y(t)$ is invariant under the transformation $t \mapsto q^{-1}t^{-1}$. It remains to show that $X(t)$ is also invariant under this transformation. We have

$$X\left(\frac{1}{qt}\right) = \sum_{[D] \in \mathcal{D}} q^{\ell(D)+g-1-\deg([D])} t^{g-1-\deg([D])}.$$

Let W be a canonical divisor of F. Then $\deg(W) = 2g - 2$ by the proof of Theorem 3.6.14, and by the same theorem

$$\ell(D) + g - 1 - \deg([D]) = \ell(W - D).$$

Therefore,

$$X\left(\frac{1}{qt}\right) = \sum_{[D] \in \mathcal{D}} q^{\ell(W-D)} t^{\deg([W-D])+1-g},$$

and the observation that $[D] \mapsto [W - D]$ is a permutation of $\mathcal{D}$ completes the proof.   □

Corollary 4.1.14. Write $L(F, t) = \sum_{i=0}^{2g} a_i t^i$ with all $a_i \in \mathbb{Z}$. Then

$$a_{2g\ i} = q^{g-i} a_i \quad \text{for } 0 \le i \le g. \tag{4.4}$$

In particular, $L(F, t)$ has degree $2g$ and leading coefficient q^g.

Proof. From the functional equation in Theorem 4.1.13 we get

$$L(F, t) = q^g t^{2g} L\left(F, \frac{1}{qt}\right) = \sum_{i=0}^{2g} a_i q^{g-i} t^{2g-i} = \sum_{i=0}^{2g} a_{2g-i} q^{i\ g} t^i.$$

A comparison of coefficients yields (4.4). Putting $i = 0$ in (4.4), we obtain $a_{2g} = q^g a_0 = q^g$ since $a_0 = 1$ by Theorem 4.1.11.   □

In view of (4.4) and $a_0 = 1$, the L-polynomial $L(F, t)$ is fully determined if we can compute the g coefficients $a_1, \ldots, a_g$. These coefficients can be conveniently calculated by a recursion, provided that the numbers $N_n(F)$, $1 \le n \le g$, of rational places of $F_n/\mathbb{F}_{q^n}$ are known.

Proposition 4.1.15. Put

$$S_n(F) = N_n(F) - q^n - 1 \quad \text{for all } n \geq 1.$$

Then

$$i\, a_i = \sum_{j=0}^{i-1} S_{i-j}(F)\, a_j \quad \text{for } 1 \leq i \leq g.$$

Proof. From Definition 4.1.4 we get

$$\frac{d}{dt} \log Z(F, t) = \sum_{n=1}^{\infty} N_n(F) t^{n-1}.$$

On the other hand, from Theorem 4.1.11 we obtain

$$\frac{d}{dt} \log Z(F, t) = \frac{d}{dt} \big(\log L(F, t) - \log(1 - t) - \log(1 - qt) \big)$$

$$= \frac{L'(F, t)}{L(F, t)} + \frac{1}{1 - t} + \frac{q}{1 - qt}$$

$$= \frac{L'(F, t)}{L(F, t)} + \sum_{n=1}^{\infty} t^{n-1} + \sum_{n=1}^{\infty} q^n t^{n-1}.$$

A comparison shows that

$$\frac{L'(F, t)}{L(F, t)} = \sum_{n=1}^{\infty} S_n(F) t^{n-1}. \tag{4.5}$$

Now writing $L(F, t)$ as in Corollary 4.1.14 and comparing the coefficients of $t^0, t^1, \ldots, t^{g-1}$ in

$$L'(F, t) = L(F, t) \sum_{n=1}^{\infty} S_n(F) t^{n-1}$$

yields the desired recursion.   □

Example 4.1.16. Let $q = 2$ and let $F = \mathbb{F}_2(x, y)$ be the Artin-Schreier extension of the rational function field $\mathbb{F}_2(x)$ defined by

$$y^2 + y = \frac{x}{x^3 + x + 1}.$$

We refer to [117, Section III.7] for the theory of Artin-Schreier extensions. The only ramified place of $\mathbb{F}_2(x)$ in the extension $F/\mathbb{F}_2(x)$ is $x^3 + x + 1$, and so F has genus $g = 2$. Therefore, the L-polynomial $L(F, t)$ of F has the form $L(F, t) = \sum_{i=0}^{4} a_i t^i$. In order to compute $L(F, t)$, we need to determine a_1 and a_2, or equivalently $N_1(F)$ and $N_2(F)$. The places ∞ and x of $\mathbb{F}_2(x)$ split completely in $F/\mathbb{F}_2(x)$, whereas the place $x + 1$ of $\mathbb{F}_2(x)$ lies under a place of F of degree 2. Thus, we have $N_1(F) = 4$. For the constant field extension $F_2 = \mathbb{F}_4(x, y)$, all five rational places of $\mathbb{F}_4(x)$ split completely in $F_2/\mathbb{F}_4(x)$, and so $N_2(F) = 10$. It follows that $S_1(F) = N_1(F) - 3 = 1$ and $S_2(F) = N_2(F) - 5 = 5$. Now we can use Proposition 4.1.15 with $a_0 = 1$ to get $a_1 = 1$ and $a_2 = 3$. The remaining coefficients of $L(F, t)$ are obtained from (4.4), namely $a_3 = q^{g-1} a_1 = 2$ and $a_4 = q^g a_0 = 4$. Therefore,

$$L(F, t) = 4t^4 + 2t^3 + 3t^2 + t + 1.$$

From this formula for $L(F, t)$ we can derive a lot of information about F; for example, we get the value of the divisor class number $h(F)$ of F by Theorem 4.1.11 as $h(F) = L(F, 1) = 11$.

4.2 The Hasse-Weil Theorem

In this section, we prove an important property of the L-polynomial $L(F, t)$ of a global function field $F/\mathbb{F}_q$, which has far-reaching consequences. As usual, let g denote the genus of F. Then, by results of the previous section, $L(F, t)$ has degree $2g$ and constant term 1. Thus, we can factor $L(F, t)$ over the complex numbers in the form

$$L(F, t) = \prod_{i=1}^{2g} (1 - \omega_i t) \qquad (4.6)$$

with $\omega_1, \ldots, \omega_{2g} \in \mathbb{C}$. Note that $\omega_1, \ldots, \omega_{2g}$ are the reciprocals of the roots of $L(F, t)$. The important fact to be established is that all ω_i, $1 \le i \le 2g$, have absolute value $q^{1/2}$. This was first proved by Weil [121], [128], whereas Hasse [52] had earlier settled the case of genus 1.

The proof proceeds in several stages. We follow a strategy suggested by Bombieri [10]. First we show that it suffices to prove a bound for the number of rational places of certain constant field extensions of F.

Lemma 4.2.1. Let $m \geq 1$ be an integer. Suppose that there exists a number C independent of n such that

$$|N_{2mn}(F) - q^{2mn} - 1| \leq Cq^{mn}$$

for all sufficiently large n. Then $|\omega_i| = q^{1/2}$ for $1 \leq i \leq 2g$.

Proof. From (4.5) and (4.6) we deduce that

$$\sum_{n=1}^{\infty} S_n(F) t^{n-1} = \frac{L'(F, t)}{L(F, t)} = -\sum_{i=1}^{2g} \frac{\omega_i}{1 - \omega_i t} = -\sum_{n=1}^{\infty} \left(\sum_{i=1}^{2g} \omega_i^n \right) t^{n-1},$$

and so a comparison of coefficients yields

$$N_n(F) - q^n - 1 = S_n(F) = -\sum_{i=1}^{2g} \omega_i^n \quad \text{for all } n \geq 1. \tag{4.7}$$

Therefore,

$$\sum_{n=1}^{\infty} S_{2mn}(F) t^n = \sum_{i=1}^{2g} \frac{\omega_i^{2m} t}{\omega_i^{2m} t - 1}.$$

By the hypothesis of the lemma, the complex power series on the left-hand side converges absolutely for $|t| < q^{-m}$, and so the rational function on the right-hand side can have no pole in the disc $|t| < q^{-m}$. This means that $|\omega_i^{2m}| \leq q^m$; hence, $|\omega_i| \leq q^{1/2}$ for $1 \leq i \leq 2g$. But $\prod_{i=1}^{2g} \omega_i = q^g$ by (4.6) and Corollary 4.1.14, thus $|\omega_i| = q^{1/2}$ for $1 \leq i \leq 2g$. □

Next we prove an upper bound on the number $N(F)$ of rational places of $F/\mathbb{F}_q$ in the case where q is a square and large relative to the genus g of F.

Lemma 4.2.2. If $F/\mathbb{F}_q$ is such that q is a square and $q > (g+1)^4$, then

$$N(F) < q + 1 + (2g+1)q^{1/2}.$$

Proof. Since the result is trivial if $N(F) = 0$, we can assume that there exists a rational place Q of F. Put $s = q^{1/2}$ and $n = s + 2g$. Let J be the set of integers j with $0 \leq j < s$ for which there exists a $u_j \in F$ with pole divisor $(u_j)_\infty = jQ$. Then the set $U = \{u_j : j \in J\}$ forms a basis of $\mathcal{L}((s-1)Q)$.

Let $\mathcal{H}$ denote the $\mathbb{F}_q$-linear subspace of F spanned by all products $x\,y^s$ with $x \in \mathcal{L}((s-1)Q)$ and $y \in \mathcal{L}(nQ)$. For any such nonzero product we have

$$\mathrm{div}(x\,y^s) = \mathrm{div}(x) + s\,\mathrm{div}(y) \geq -(s-1)Q - sn\,Q,$$

and so $\mathcal{H} \subseteq \mathcal{L}((s-1+sn)Q)$. Since U is a basis of $\mathcal{L}((s-1)Q)$ and s is a power of the characteristic of F, any $w \in \mathcal{H}$ can be written in the form

$$w = \sum_{j \in J} u_j w_j^s \tag{4.8}$$

with the u_j as above and all $w_j \in \mathcal{L}(nQ)$. In fact, this representation is unique. Suppose, on the contrary, that

$$\sum_{j \in J} u_j y_j^s = 0 \tag{4.9}$$

with all $y_j \in \mathcal{L}(nQ)$ and not all $y_j = 0$. For any $j \in J \subseteq \{0, 1, \ldots, s-1\}$ with $y_j \neq 0$ we have

$$v_Q(u_j y_j^s) \equiv v_Q(u_j) \equiv -j \pmod{s},$$

and so the numbers on the left-hand side of this congruence are distinct. Therefore, the strict triangle inequality (see Remark 1.5.4(i)) yields

$$v_Q\left(\sum_{j \in J} u_j y_j^s\right) = \min_{j \in J}\, v_Q(u_j y_j^s) \neq \infty,$$

a contradiction to (4.9). The uniqueness of the representation (4.8) and Riemann's theorem show that

$$\dim(\mathcal{H}) = \ell((s-1)Q)\,\ell(nQ) \geq (s-g)(n+1-g).$$

Since $q > (g+1)^4$, we obtain

$$(s-g)(n+1-g) = (s-g)(s+g+1) = q - g^2 + s - g$$
$$= q + g + 1 + s - (g+1)^2 > q + g + 1,$$

and so by the Riemann-Roch theorem

$$\dim(\mathcal{H}) > q + g + 1 = \ell((q + 2g)Q). \tag{4.10}$$

Now we set up an additive group homomorphism $\psi : \mathcal{H} \rightarrow \mathcal{L}((s^2 - s + n)Q) = \mathcal{L}((q + 2g)Q)$ by taking $w \in \mathcal{H}$ in the form (4.8) and putting

$$\psi(w) = \sum_{j \in J} u_j^s w_j.$$

By (4.10) there exists an element $z \neq 0$ in the kernel of ψ. Write z in the form (4.8), say

$$z = \sum_{j \in J} u_j z_j^s$$

with all $z_j \in \mathcal{L}(nQ)$.

Consider any rational place $P \neq Q$ of F. Note that Q is the only possible pole of the u_j and z_j, and so these elements, and also z, are in $\mathcal{O}_P$. Hence, it makes sense to look at their residue classes relative to P (see Definition 1.5.10). Since the residue class field of P is $\mathbb{F}_q$, we have

$$z(P)^s = \left(\sum_{j \in J} u_j(P) z_j(P)^s \right)^s = \sum_{j \in J} u_j(P)^s z_j(P)^q$$

$$= \left(\sum_{j \in J} u_j^s z_j \right)(P) = \psi(z)(P) = 0,$$

and so $z(P) = 0$. Thus, any $P \neq Q$ is a zero of z, and hence the zero divisor $(z)_0$ satisfies

$$\deg((z)_0) \geq N(F) - 1.$$

On the other hand, we have $z \in \mathcal{H} \subseteq \mathcal{L}((s - 1 + sn)Q)$ and this implies by Corollary 3.4.3,

$$\deg((z)_0) = \deg((z)_\infty) \leq s - 1 + sn = q + (2g + 1)q^{1/2} - 1.$$

Combining these two inequalities, we get the bound in the lemma. $\square$

We can now establish the final result we are aiming at. We use Galois coverings of curves (see Section 3.8) to derive the bound that is needed in Lemma 4.2.1 from the preliminary bound in Lemma 4.2.2.

Theorem 4.2.3 (Hasse-Weil Theorem). Let

$$L(F, t) = \prod_{i=1}^{2g}(1 - \omega_i t)$$

be the L-polynomial of the global function field $F/\mathbb{F}_q$ of genus g. Then $|\omega_i| = q^{1/2}$ for $1 \le i \le 2g$.

Proof. We can choose $x \in F \backslash \mathbb{F}_q$ such that $F/\mathbb{F}_q(x)$ is a finite separable extension (see [51, p. 27] and [117, Proposition III.9.2]). Let $K/\mathbb{F}_q(x)$ be the Galois closure of $F/\mathbb{F}_q(x)$, that is, K is the smallest algebraic extension of F (in a fixed algebraic closure of F) that is Galois over $\mathbb{F}_q(x)$. If $\mathbb{F}_{q^m}$ is the full constant field of K, then we consider the tower $\mathbb{F}_{q^m}(x) \subseteq F_m = F \cdot \mathbb{F}_{q^m} \subseteq K$ in which the extension $K/\mathbb{F}_{q^m}(x)$ is again Galois. In view of Lemma 4.2.1, it suffices to consider the number of rational places in constant field extensions of F_m. Thus, we may change notation and assume from now on that $\mathbb{F}_q$ is also the full constant field of K.

For a positive integer n, we consider now the tower $\mathbb{F}_{q^{2n}}(x) \subseteq F_{2n} \subseteq K_{2n} := K \cdot \mathbb{F}_{q^{2n}}$ and put $k_n = \mathbb{F}_{q^{2n}}$, $G_n = \mathrm{Gal}(K_{2n}/k_n(x))$, and $H_n = \mathrm{Gal}(K_{2n}/F_{2n})$. We observe that $|G_n| = [K_{2n} : k_n(x)] \le [K : \mathbb{F}_q(x)]$ is bounded uniformly in n. By turning to the corresponding nonsingular projective curves, we get $\mathcal{A}_n \to \mathcal{B}_n \to \mathbf{P}^1(\overline{\mathbb{F}_q})$, where $\mathcal{A}_n$ is a Galois covering of $\mathbf{P}^1(\overline{\mathbb{F}_q})$ with Galois group G_n and $\mathcal{A}_n$ is a Galois covering of $\mathcal{B}_n$ with Galois group $H_n \subseteq G_n$. Note that the k_n-rational points of $\mathcal{A}_n$ and $\mathcal{B}_n$ are lying over the k_n-rational points of $\mathbf{P}^1(\overline{\mathbb{F}_q})$.

Denote by R_n the preimage in $\mathcal{A}_n$ of the k_n-rational points of $\mathbf{P}^1(\overline{\mathbb{F}_q})$. For any $p \in \mathbf{P}^1(\overline{\mathbb{F}_q})$, the Galois group G_n acts transitively on the fiber of p in $\mathcal{A}_n$ and the Frobenius morphism ϕ on $\mathcal{A}_n$ also acts on the fiber of p. Given r in the fiber of p, we have $\phi(r) = \sigma(r)$ for some $\sigma \in G_n$. If the covering $\mathcal{A}_n \to \mathbf{P}^1(\overline{\mathbb{F}_q})$ is unramified at p, then there are $|G_n|$ points in the fiber of p and σ is unique. For any $\sigma \in G_n$, we put

$$R_n(\sigma) := \{r \in R_n : \phi(r) = \sigma(r)\}.$$

By a similar argument as in the proof of Lemma 4.2.2, but using compositions with ϕ and σ, we obtain the bound

$$|R_n(\sigma)| < q^{2n} + 1 + (2g' + 1)q^n$$

for all sufficiently large n, where g' is the genus of $\mathcal{A}_n$ which is independent of n by [117, Theorem III.6.3(b)]. A counting argument yields

$$\sum_{\sigma \in G_n} |R_n(\sigma)| = |G_n|(q^{2n} + 1) + O(1),$$

where $O(1)$ takes care of the ramified points in the covering $\mathcal{A}_n \to \mathbf{P}^1(\overline{\mathbb{F}_q})$. In view of the upper bound on $|R_n(\sigma)|$ for all $\sigma \in G_n$, this implies

$$|R_n(\sigma)| \geq q^{2n} - (|G_n| - 1)(2g' + 1)q^n + O(1)$$

for all $\sigma \in G_n$ and all sufficiently large n, and so

$$|R_n(\sigma)| = q^{2n} + O(q^n) \quad \text{for all } \sigma \in G_n, \tag{4.11}$$

with an implied constant independent of n.

The points of $\mathcal{B}_n$ are the H_n-orbits of the points of $\mathcal{A}_n$ and the k_n-rational points of $\mathcal{B}_n$ are those orbits $H_n r$ such that $H_n r = (H_n \phi)r$, that is, $\phi(r) = \sigma(r)$ for some $\sigma \in H_n$. These points r belong to R_n. Therefore, the set of points of $\mathcal{A}_n$ lying over the k_n-rational points of $\mathcal{B}_n$ is given by $\cup_{\sigma \in H_n} R_n(\sigma)$. Again by a counting argument, we obtain

$$\sum_{\sigma \in H_n} |R_n(\sigma)| = |H_n| N_{2n}(F) + O(1),$$

and so (4.11) yields

$$N_{2n}(F) = q^{2n} + O(q^n)$$

with an implied constant independent of n. Thus, the hypothesis of Lemma 4.2.1 is satisfied and we get the result of the theorem. □

We derive two consequences from the Hasse-Weil theorem. The first one is a fundamental bound on the number of rational places of a global function field or, equivalently, on the number of $\mathbb{F}_q$-rational points of a nonsingular projective curve over $\mathbb{F}_q$.

Theorem 4.2.4 (Hasse-Weil Bound). Let $F/\mathbb{F}_q$ be a global function field of genus g. Then the number $N(F)$ of rational places of F satisfies

$$|N(F) - q - 1| \leq 2g q^{1/2}.$$

Proof. From (4.7) with $n = 1$ and Theorem 4.2.3 we obtain

$$|N(F) - q - 1| = \left| \sum_{i=1}^{2g} \omega_i \right| \le \sum_{i=1}^{2g} |\omega_i| = 2gq^{1/2}.$$

$\square$

Theorem 4.2.5. The divisor class number $h(F)$ of a global function field $F/\mathbb{F}_q$ of genus g satisfies

$$(q^{1/2} - 1)^{2g} \le h(F) \le (q^{1/2} + 1)^{2g}.$$

Proof. The last part of Theorem 4.1.11 yields

$$h(F) = L(F, 1) = \prod_{i=1}^{2g} |\omega_i - 1|,$$

and so an application of Theorem 4.2.3 leads to the desired bounds. $\square$

Example 4.2.6. Let q be the square of a prime power and let $F/\mathbb{F}_q$ be a global function field of genus $g \ge 1$ with $N(F) = q + 1 + 2gq^{1/2}$. Note that this is the largest value of $N(F)$ that is allowed by the Hasse-Weil bound in Theorem 4.2.4 for the given values of q and g. Now we consider the identity

$$N(F) = q + 1 - \sum_{i=1}^{2g} \omega_i,$$

which is obtained from (4.7) with $n = 1$. We observe that since $|\omega_i| = q^{1/2}$ for $1 \le i \le 2g$ by Theorem 4.2.3, the above identity for $N(F)$ can hold only if $\omega_i = -q^{1/2}$ for $1 \le i < 2g$. It follows then that the L-polynomial $L(F, t)$ of F is given by $L(F, t) = (1 + q^{1/2}t)^{2g}$. In particular, for the class number $h(F)$ of F we obtain $h(F) = (1 + q^{1/2})^{2g}$ by Theorem 4.1.11. An example of a global function field $F/\mathbb{F}_q$, which satisfies the condition $N(F) = q + 1 + 2gq^{1/2}$, is provided by the Hermitian function field H_q over $\mathbb{F}_q$, since in this case we have $g(H_q) = (q - q^{1/2})/2$ and $N(H_q) = q^{3/2} + 1$ (see [117, Lemma VI.4.4]). If $q = r^2$ with a prime power r, then H_q is given by $H_q = \mathbb{F}_q(x, y)$ with $y^r + y = x^{r+1}$.

4.3 Further Bounds and Asymptotic Results

The Hasse-Weil bound in Theorem 4.2.4 can be refined in various cases. For
instance, the following bound due to Serre [112] is of interest in the case
where q is not a square since it may then yield an improvement on the Hasse-
Weil bound. Here and in the following, we write $\lfloor u \rfloor$ for the floor function of
a real number u, that is, for the greatest integer $\leq u$.

> **Theorem 4.3.1 (Serre Bound).** Let $F/\mathbb{F}_q$ be a global function field of
> genus g. Then the number $N(F)$ of rational places of F satisfies
>
> $$|N(F) - q - 1| \leq g\lfloor 2q^{1/2} \rfloor.$$

Proof. We can assume that $g > 0$. Let $\omega_1, \ldots, \omega_{2g}$ be the reciprocals
of the roots of the L-polynomial $L(F, t)$ of F. We claim that we
can pair off and arrange these complex numbers in such a way that
$\omega_{g+i} = \overline{\omega_i}$ for $1 \leq i \leq g$. This is clear if ω_i is not real. By
Theorem 4.2.3, the only possible real values among $\omega_1, \ldots, \omega_{2g}$ are
$\pm q^{1/2}$. Now (4.6) and Corollary 4.1.14 imply that $\prod_{i=1}^{2g} \omega_i = q^g > 0$;
hence, the value $-q^{1/2}$ must appear an even number of times among
$\omega_1, \ldots, \omega_{2g}$, and so the desired pairing off is again possible. Finally,
since the total number of the ω_i, $1 \leq i \leq 2g$, is $2g$ and thus even,
the remaining value $q^{1/2}$ appears also an even number of times
among them.

Given this arrangement of $\omega_1, \ldots, \omega_{2g}$, we put

$$\gamma_i := \omega_i + \omega_{g+i} + m + 1 = \omega_i + \overline{\omega_i} + m + 1 \quad \text{for } 1 \leq i \leq g,$$

where $m = \lfloor 2q^{1/2} \rfloor$. Then

$$\gamma_i \geq m + 1 - |\omega_i + \overline{\omega_i}| \geq m + 1 - 2q^{1/2} > 0 \quad \text{for } 1 \leq i \leq g.$$

Since $\omega_1, \ldots, \omega_{2g}$ are exactly all roots of the monic polynomial
$t^{2g}L(F, 1/t) \in \mathbb{Z}[t]$, they are algebraic integers, and so $\prod_{i=1}^{g} \gamma_i > 0$
is an integer. Therefore,

$$\prod_{i=1}^{g} \gamma_i \geq 1.$$

It follows then from (4.7) with $n = 1$ that

$$1 + m - \frac{1}{g}(N(F) - q - 1) = \frac{1}{g}\left(g + gm + \sum_{i=1}^{g}(\omega_i + \omega_{g+i})\right)$$

$$= \frac{1}{g}\sum_{i=1}^{g}\gamma_i \geq \left(\prod_{i=1}^{g}\gamma_i\right)^{1/g} \geq 1,$$

and this yields the upper bound

$$N(F) \leq q + 1 + gm.$$

To prove the corresponding lower bound, we work with the numbers $\delta_i := m + 1 - (\omega_i + \overline{\omega_i})$, $1 \leq i \leq g$, instead of the γ_i and obtain

$$1 + m + \frac{1}{g}(N(F) - q - 1) = \frac{1}{g}\sum_{i=1}^{g}\delta_i \geq 1;$$

hence,

$$N(F) \geq q + 1 - gm,$$

and the proof is complete. $\square$

From the Hasse-Weil and Serre bounds we see that the number of rational places of a global function field $F/\mathbb{F}_q$ of genus g is bounded from above by a number depending only on q and g. Hence, the following definition makes sense.

Definition 4.3.2. For a given prime power q and an integer $g \geq 0$, let $N_q(g)$ denote the maximum number of rational places that a global function field $F/\mathbb{F}_q$ of genus g can have.

Note that $N_q(g)$ is also the maximum number of $\mathbb{F}_q$-rational points that a nonsingular projective curve over $\mathbb{F}_q$ of genus g can have. It is trivial that $N_q(0) = q + 1$. From the Serre bound we obtain

$$N_q(g) \leq q + 1 + g\lfloor 2q^{1/2} \rfloor \tag{4.12}$$

for all q and g. If g is relatively large with respect to q, then the following result provides an improved bound.

Theorem 4.3.3. For an integer $r \geq 1$, suppose that $c_1, \ldots, c_r$ are r nonnegative real numbers such that at least one of them is nonzero and the inequality

$$1 + \lambda_r(t) + \lambda_r(t^{-1}) \geq 0$$

holds for all $t \in \mathbb{C}$ with $|t| = 1$, where $\lambda_r(t) = \sum_{n=1}^{r} c_n t^n$. Then we have

$$N_q(g) \leq \frac{g}{\lambda_r(q^{-1/2})} + \frac{\lambda_r(q^{1/2})}{\lambda_r(q^{-1/2})} + 1$$

for all q and g.

Proof. The bound is trivial for $g = 0$ since $\lambda_r(q^{1/2}) \geq q\,\lambda_r(q^{-1/2})$, and so we can assume that $g > 0$. Let $F/\mathbb{F}_q$ be an arbitrary global function field of genus g and pair off and arrange $\omega_1, \ldots, \omega_{2g}$ as in the proof of Theorem 4.3.1. With $\alpha_i = \omega_i q^{-1/2}$ we have $|\alpha_i| = 1$ for $1 \leq i \leq 2g$ and $\alpha_{g+i} = \alpha_i^{-1}$ for $1 \leq i \leq g$. Using the trivial bound $N(F) \leq N_n(F)$ and (4.7), we obtain

$$N(F)q^{-n/2} \leq N_n(F)q^{-n/2} = q^{n/2} + q^{-n/2} - \sum_{i=1}^{g}(\alpha_i^n + \alpha_i^{-n})$$

for any positive integer n. By forming an appropriate linear combination of these inequalities, we get

$$N(F)\lambda_r(q^{-1/2}) \leq \lambda_r(q^{1/2}) + \lambda_r(q^{-1/2}) - \sum_{i=1}^{g}(\lambda_r(\alpha_i) + \lambda_r(\alpha_i^{-1}))$$

$$\leq \lambda_r(q^{1/2}) + \lambda_r(q^{-1/2}) + g,$$

where we used the condition on λ_r in the second step. Now the desired result follows from the definition of $N_q(g)$. $\qquad\square$

Example 4.3.4. Here is a concrete application of Theorem 4.3.3. Take $q = 2$, $r = 6$, and

$$c_1 = \frac{184}{203}, \ c_2 = \frac{20}{29}, \ c_3 = \frac{90}{203}, \ c_4 = \frac{89}{406}, \ c_5 = \frac{2}{29}, \ c_6 = \frac{2}{203}.$$

Then the condition on λ_6 in Theorem 4.3.3 is satisfied since it is easily checked that

$$1 + \lambda_6(t) + \lambda_6(t^{-1}) = \frac{1}{406} \left(f(t) + f(t^{-1}) \right)^2$$

with

$$f(t) = 2t^3 + 7t^2 + 10t + 5.$$

After a simple calculation, Theorem 4.3.3 yields the bound

$$N_2(g) \leq (0.83)\, g + 5.35 \quad \text{for all } g \geq 0.$$

Example 4.3.5. Take $q = 2$ and $g = 5$. Then the bound in Example 4.3.4 yields $N_2(5) \leq 9$. This bound is best possible. To see this, consider the following example. Let $F = \mathbb{F}_2(x, y_1, y_2)$ with

$$y_1^2 + y_1 = x^3 + x, \qquad y_2^2 + y_2 = (x^2 + x)\, y_1.$$

Then F has genus 5 by the theory of Artin-Schreier extensions (see [117, Section III.7]). Furthermore, we have $N(F) = 9$ since there are four rational places of F lying over each of the places x and $x + 1$ of $\mathbb{F}_2(x)$ and there is a totally ramified place lying over the infinite place of $\mathbb{F}_2(x)$. Thus, we have shown that $N_2(5) = 9$. Note that the Serre bound (4.12) yields only $N_2(5) \leq 13$, whereas the Hasse-Weil bound in Theorem 4.2.4 yields the even worse bound $N_2(5) \leq 17$.

Tables of values of, or bounds on, $N_q(g)$ can be found in van der Geer and van der Vlugt [41] and in the book of Niederreiter and Xing [100, Section 4.5].

It is of great interest to study the asymptotic behavior of $N_q(g)$ for fixed q and $g \to \infty$. Since the upper bounds on $N_q(g)$ that we have established grow linearly with g for fixed q, it is reasonable to introduce the following quantity.

Definition 4.3.6. For any prime power q, define

$$A(q) = \limsup_{g \to \infty} \frac{N_q(g)}{g}.$$

It follows from the Serre bound (4.12) that $A(q) \leq \lfloor 2q^{1/2} \rfloor$ for all q. Example 4.3.4 indicates that improvements are possible, since the upper bound on $N_2(g)$ in this example implies that $A(2) \leq 0.83$. By making a better use of Theorem 4.3.3, the following significant improvement was obtained by Vlăduţ and Drinfeld [125].

Theorem 4.3.7. For every prime power q, we have

$$A(q) \leq q^{1/2} - 1.$$

Proof. For any positive integer r, consider the polynomial

$$\lambda_r(t) = \sum_{n=1}^{r} \left(1 - \frac{n}{r+1} \right) t^n.$$

For any $t \in \mathbb{C}$ with $|t| = 1$, we have

$$1 + \lambda_r(t) + \lambda_r(t^{-1}) = \sum_{n=-r}^{r} \left(1 - \frac{|n|}{r+1} \right) t^n = \frac{1}{r+1} \sum_{j,k=0}^{r} t^{j-k}$$

$$= \frac{1}{r+1} \left| \sum_{n=0}^{r} t^n \right|^2 \geq 0,$$

and so λ_r can be used in Theorem 4.3.3. For $g > 0$ this yields

$$\frac{N_q(g)}{g} \leq \frac{1}{\lambda_r(q^{-1/2})} + \frac{\lambda_r(q^{1/2})}{\lambda_r(q^{-1/2}) g} + \frac{1}{g}. \tag{4.13}$$

If $0 \leq t < 1$, then

$$\sum_{n=1}^{r} t^n \geq \lambda_r(t) \geq \sum_{n=1}^{r} t^n - \frac{1}{r+1} \sum_{n=1}^{\infty} n \, t^n,$$

and so

$$\lim_{r \to \infty} \lambda_r(q^{-1/2}) = \sum_{n=1}^{\infty} q^{-n/2} = \frac{1}{q^{1/2} - 1}.$$

Furthermore,

$$\lambda_r(q^{1/2}) \leq \sum_{n=1}^{r} q^{n/2} < \frac{q^{(r+1)/2}}{q^{1/2} - 1}.$$

For $g \geq q$ we choose $r = r(g)$ to be the largest integer with $q^r \leq g$. Then

$$\lambda_r(q^{1/2}) < \frac{q^{1/2} g^{1/2}}{q^{1/2} - 1}.$$

Now we let g tend to infinity in (4.13) in such a way that $N_q(g)/g \rightarrow A(q)$. Then with our choice of r we have $r \rightarrow \infty$, and so we obtain the desired bound on $A(q)$. □

Remark 4.3.8. The bound on $A(q)$ in Theorem 4.3.7 is best possible in the case where q is a square, since then it is known that $A(q) = q^{1/2} - 1$ (see Ihara [59] and Garcia and Stichtenoth [36, 37]). For prime powers q that are not squares, so in particular for primes q, the value of $A(q)$ is unknown. However, a general lower bound on $A(q)$ is available, namely

$$A(q) > c \log q$$

for all prime powers q, with an absolute constant $c > 0$. A possible value of c is $c = (96 \log 2)^{-1}$ (see Niederreiter and Xing [100, Theorem 5.2.9]). For further lower bounds on $A(q)$, we refer to Li and Maharaj [69], Niederreiter and Xing [99], and Temkine [120], as well as to the recent survey articles of Garcia and Stichtenoth [38] and Maharaj [77], which discuss also constructive aspects. In the interesting case $q = 2$, a known lower bound is $A(2) \geq \frac{81}{317} = 0.2555\ldots$ (see Niederreiter and Xing [99]) and this was recently improved to $A(2) \geq \frac{97}{376} = 0.2579\ldots$ by Xing and Yeo [136]. This should be compared with the upper bound $A(2) \leq \sqrt{2} - 1 = 0.4142\ldots$ obtained from Theorem 4.3.7.

4.4 Character Sums

Character sums over finite fields arise in various applications to number theory, coding theory, and other areas. The Hasse-Weil bound in Theorem 4.2.4 has important implications for character sums with polynomial arguments, in the sense that it leads to very useful upper bounds on such character sums.

Before establishing these results, we recall the fundamentals of character theory for finite abelian groups and finite fields.

For a given finite abelian group G, a *character* χ of G is a group homomorphism from G into the multiplicative group of complex numbers of absolute value 1. Among the characters of G we have the *trivial character* χ_0 defined by $\chi_0(c) = 1$ for all $c \in G$; all other characters of G are called *nontrivial*. The number of characters of G is equal to the order of G. We will make use of the following simple fact which is easily proved.

Lemma 4.4.1. If χ is a nontrivial character of the finite abelian group G, then

$$\sum_{c \in G} \chi(c) = 0.$$

Proof. Suppose, for the sake of concreteness, that G is written multiplicatively. Since χ is nontrivial, there exists $b \in G$ with $\chi(b) \neq 1$. Then

$$\chi(b) \sum_{c \in G} \chi(c) = \sum_{c \in G} \chi(bc) = \sum_{c \in G} \chi(c),$$

for if c runs through G, then so does bc. Hence, we have

$$(\chi(b) - 1) \sum_{c \in G} \chi(c) = 0,$$

which implies the desired result. $\qquad\qquad\qquad\qquad\qquad\qquad$ □

In a finite field $\mathbb{F}_q$ there are two finite abelian groups of structural significance, namely the additive group of $\mathbb{F}_q$ and the multiplicative group $\mathbb{F}_q^*$ of nonzero elements of $\mathbb{F}_q$. We make a corresponding distinction between the characters pertaining to these groups, by speaking of additive and multiplicative characters of $\mathbb{F}_q$.

An *additive character* of $\mathbb{F}_q$ is a character of the additive group of $\mathbb{F}_q$. Thus, such a character χ satisfies

$$\chi(b + c) = \chi(b)\chi(c) \quad \text{for all } b, c \in \mathbb{F}_q.$$

The additive characters of $\mathbb{F}_q$ can be described as follows. Let p be the characteristic of $\mathbb{F}_q$ and identify the prime field $\mathbb{F}_p$ contained in $\mathbb{F}_q$ with

$\mathbb{Z}/p\mathbb{Z}$. Let $\mathrm{Tr} : \mathbb{F}_q \to \mathbb{F}_p$ be the trace map from $\mathbb{F}_q$ to $\mathbb{F}_p$ (see Section 1.4). Now fix $a \in \mathbb{F}_q$ and define

$$\chi_a(c) = e^{2\pi i \, \mathrm{Tr}(ac)/p} \quad \text{for all } c \in \mathbb{F}_q, \tag{4.14}$$

where $i = \sqrt{-1}$. It is clear that χ_a is an additive character of $\mathbb{F}_q$. By letting a run through $\mathbb{F}_q$, we get exactly all the q additive characters of $\mathbb{F}_q$.

A *multiplicative character* of $\mathbb{F}_q$ is a character of the multiplicative group $\mathbb{F}_q^*$. Thus, such a character ψ satisfies

$$\psi(bc) = \psi(b)\,\psi(c) \quad \text{for all } b, c \in \mathbb{F}_q^*.$$

Since $\mathbb{F}_q^*$ is a cyclic group, its characters are easily described. Fix a primitive element r of $\mathbb{F}_q$ (see Definition 1.1.7). Then for each integer $j = 0, 1, \ldots, q - 2$, the function ψ_j defined by

$$\psi_j(r^k) = e^{2\pi i j k/(q-1)} \quad \text{for } k = 0, 1, \ldots, q - 2 \tag{4.15}$$

is a multiplicative character of $\mathbb{F}_q$, and all the $q - 1$ multiplicative characters of $\mathbb{F}_q$ are obtained in this way.

As a preparation for the main results of this section, we develop a useful general principle. Let M_q denote the set of all monic polynomials over $\mathbb{F}_q$. Let λ be a complex-valued function on M_q, which satisfies $|\lambda(g)| \leq 1$ for all $g \in M_q$, $\lambda(1) = 1$, and

$$\lambda(gh) = \lambda(g)\lambda(h) \quad \text{for all } g, h \in M_q. \tag{4.16}$$

For any integer $n \geq 0$, let $M_q^{(n)}$ denote the set of all monic polynomials over $\mathbb{F}_q$ of degree n.

Lemma 4.4.2. Suppose that for some integer $m \geq 1$ we have

$$\sum_{g \in M_q^{(n)}} \lambda(g) = 0 \quad \text{for all } n > m. \tag{4.17}$$

Then there exist complex numbers $\beta_1, \ldots, \beta_m$ such that

$$L_s := \sum_P \deg(P)\,\lambda(P)^{s/\deg(P)} = -\sum_{l=1}^{m} \beta_l^s \quad \text{for all } s \geq 1,$$

where the first sum is extended over all monic irreducible polynomials P in $\mathbb{F}_q[x]$ with $\deg(P)$ dividing s.

Proof. By the hypothesis (4.17),

$$L(t) := \sum_{n=0}^{\infty} \left(\sum_{g \in M_q^{(n)}} \lambda(g) \right) t^n$$

is a complex polynomial of degree at most m with constant term 1. Hence, we can write

$$L(t) = \prod_{l=1}^{m} (1 - \beta_l t)$$

with some complex numbers $\beta_1, \ldots, \beta_m$. It follows that

$$t \frac{d \log L(t)}{dt} = -\sum_{l=1}^{m} \frac{\beta_l t}{1 - \beta_l t} = -\sum_{s=1}^{\infty} \left(\sum_{l=1}^{m} \beta_l^s \right) t^s \qquad (4.18)$$

for sufficiently small $|t|$. On the other hand, because of (4.16) and unique factorization in $\mathbb{F}_q[x]$, we can write for $|t| < 1$,

$$L(t) = \sum_{g \in M_q} \lambda(g)\, t^{\deg(g)} = \prod_{P} (1 + \lambda(P)\, t^{\deg(P)} + \lambda(P^2) t^{\deg(P^2)} + \cdots)$$

$$= \prod_{P} (1 + \lambda(P)\, t^{\deg(P)} + \lambda(P)^2 t^{2 \deg(P)} + \cdots)$$

$$= \prod_{P} (1 - \lambda(P)\, t^{\deg(P)})^{-1},$$

where the products are taken over all monic irreducible polynomials P in $\mathbb{F}_q[x]$. This implies

$$t \frac{d \log L(t)}{dt} = \sum_{P} \frac{\lambda(P) \deg(P)\, t^{\deg(P)}}{1 - \lambda(P)\, t^{\deg(P)}}$$

$$= \sum_{P} \lambda(P) \deg(P)\, t^{\deg(P)} (1 + \lambda(P)\, t^{\deg(P)} + \lambda(P)^2\, t^{2 \deg(P)} + \cdots)$$

$$= \sum_{P} \deg(P)(\lambda(P)\, t^{\deg(P)} + \lambda(P)^2 t^{2 \deg(P)} + \lambda(P)^3 t^{3 \deg(P)} + \cdots).$$

Collecting equal powers of t, we obtain

$$t \frac{d \log L(t)}{dt} = \sum_{s=1}^{\infty} L_s t^s,$$

and a comparison with (4.18) yields the desired result.   □

Given a polynomial $f \in \mathbb{F}_q[x]$ and a nontrivial additive character χ of $\mathbb{F}_q$, we now consider the character sum

$$\sum_{c \in \mathbb{F}_q} \chi(f(c)).$$

Since the case where $\deg(f) \leq 1$ is trivial, we will assume in the following two results that $d := \deg(f) \geq 2$. We start by establishing a simple algebraic fact.

Lemma 4.4.3. Let $f \in \mathbb{F}_q[x]$ be of degree $d \geq 2$ with leading coefficient b_d. For positive integers $r \leq n$, let $\sigma_r = \sigma_r(x_1, \ldots, x_n)$ be the rth elementary symmetric polynomial in the indeterminates $x_1, \ldots, x_n$ over $\mathbb{F}_q$. Then for $n \geq d$ we have

$$f(x_1) + \cdots + f(x_n) = (-1)^{d-1} d\, b_d\, \sigma_d + G(\sigma_1, \ldots, \sigma_{d-1})$$

for some polynomial G in $d - 1$ indeterminates over $\mathbb{F}_q$. For $1 \leq n < d$ we have

$$f(x_1) + \cdots + f(x_n) = H(\sigma_1, \ldots, \sigma_n)$$

for some polynomial H in n indeterminates over $\mathbb{F}_q$.

Proof. Fix $n \geq 1$. For $j \geq 1$ put

$$w_j(x_1, \ldots, x_n) := x_1^j + \cdots + x_n^j \in \mathbb{F}_q[x_1, \ldots, x_n].$$

By Waring's formula we have

$$w_j(x_1, \ldots, x_n) = \sum (-1)^{i_2 + i_4 + i_6 + \cdots} \frac{(i_1 + i_2 + \cdots + i_n - 1)!\, j}{i_1!\, i_2! \cdots i_n!} \sigma_1^{i_1} \sigma_2^{i_2} \cdots \sigma_n^{i_n},$$

where the summation is extended over all n-tuples $(i_1, \ldots, i_n)$ of nonnegative integers with $i_1 + 2i_2 + \cdots + ni_n = j$. For $j = 1$ we get $w_1(x_1, \ldots, x_n) = \sigma_1$. For $2 \le j \le n$, there is one solution of $i_1 + 2i_2 + \cdots + ni_n = j$ with $i_j = 1$ and with all other $i_h = 0$, and the term corresponding to this solution is $(-1)^{j-1} j\sigma_j$. All other solutions of $i_1 + 2i_2 + \cdots + ni_n = j$ have $i_j = i_{j+1} = \cdots = i_n = 0$, and so the corresponding terms involve only $\sigma_1, \ldots, \sigma_{j-1}$. Thus,

$$w_1(x_1, \ldots, x_n) = \sigma_1,$$
$$w_j(x_1, \ldots, x_n) = (-1)^{j-1} j\sigma_j + G_j(\sigma_1, \ldots, \sigma_{j-1}) \quad \text{for } 2 \le j \le n,$$
$$w_j(x_1, \ldots, x_n) = H_j(\sigma_1, \ldots, \sigma_n) \quad \text{for } j > n,$$

where G_j is a polynomial in $j - 1$ and H_j a polynomial in n indeterminates over $\mathbb{F}_q$. If we write

$$f(x) = b_d x^d + \cdots + b_1 x + b_0$$

with all $b_i \in \mathbb{F}_q$, then

$$f(x_1) + \cdots + f(x_n) = \sum_{j=1}^{d} b_j w_j(x_1, \ldots, x_n) + n b_0,$$

and the desired result follows. $\qquad\square$

A given additive character χ of $\mathbb{F}_q$ can be lifted canonically to an additive character $\chi^{(s)}$ of an extension field $\mathbb{F}_{q^s}$ of $\mathbb{F}_q$. This is done by using the trace map $\mathrm{Tr}_s : \mathbb{F}_{q^s} \to \mathbb{F}_q$ from $\mathbb{F}_{q^s}$ to $\mathbb{F}_q$ and defining

$$\chi^{(s)}(\gamma) = \chi(\mathrm{Tr}_s(\gamma)) \quad \text{for all } \gamma \in \mathbb{F}_{q^s}. \tag{4.19}$$

Proposition 4.4.4. Let $f \in \mathbb{F}_q[x]$ be of degree $d \ge 2$ with $\gcd(d, q) = 1$ and let χ be a nontrivial additive character of $\mathbb{F}_q$. Then there exist complex numbers $\beta_1, \ldots, \beta_{d-1}$, depending only on f and χ, such that for any integer $s \ge 1$ we have

$$\sum_{\gamma \in \mathbb{F}_{q^s}} \chi^{(s)}(f(\gamma)) = -\sum_{l=1}^{d-1} \beta_l^s.$$

Proof. We define a complex-valued function λ on M_q as follows. We put $\lambda(1) = 1$. If $g \in M_q^{(n)}$ for some $n \ge 1$, then let $g(x) = (x - \alpha_1) \cdots (x - \alpha_n)$ be the factorization of g in its splitting field over $\mathbb{F}_q$. Since $\sigma_r(\alpha_1, \ldots, \alpha_n) \in \mathbb{F}_q$ for $1 \le r \le n$, it follows from Lemma 4.4.3 that $f(\alpha_1) + \cdots + f(\alpha_n) \in \mathbb{F}_q$. We put

$$\lambda(g) = \chi\big(f(\alpha_1) + \cdots + f(\alpha_n)\big).$$

If $h(x) = (x - \rho_1) \cdots (x - \rho_k) \in M_q$, then

$$\lambda(gh) = \chi\big(f(\alpha_1) + \cdots + f(\alpha_n) + f(\rho_1) + \cdots + f(\rho_k)\big)$$
$$= \chi\big(f(\alpha_1) + \cdots + f(\alpha_n)\big)\chi\big(f(\rho_1) + \cdots + f(\rho_k)\big) = \lambda(g)\lambda(h),$$

and so (4.16) is satisfied.

Next we verify (4.17) with $m = d - 1$. Thus, we consider the sum

$$\sum_{g \in M_q^{(n)}} \lambda(g)$$

for fixed $n \ge d$. For

$$g(x) = x^n + \sum_{r=1}^{n} (-1)^r a_r x^{n-r} = (x - \alpha_1) \cdots (x - \alpha_n) \in M_q^{(n)}$$

we have $\sigma_r(\alpha_1, \ldots, \alpha_n) = a_r$ for $1 \le r \le n$, and so Lemma 4.4.3 implies that

$$f(\alpha_1) + \cdots + f(\alpha_n) = (-1)^{d-1} d\, b_d a_d + G(a_1, \ldots, a_{d-1}).$$

Since $\gcd(d, q) = 1$, we have $b := (-1)^{d-1} d\, b_d \ne 0$; hence,

$$\sum_{g \in M_q^{(n)}} \lambda(g) = \sum_{a_1, \ldots, a_n \in \mathbb{F}_q} \chi(b\, a_d + G(a_1, \ldots, a_{d-1}))$$
$$= q^{n-d} \sum_{a_1, \ldots, a_d \in \mathbb{F}_q} \chi(b\, a_d)\chi(G(a_1, \ldots, a_{d-1}))$$
$$= q^{n-d} \left(\sum_{a_d \in \mathbb{F}_q} \chi(b\, a_d) \right) \left(\sum_{a_1, \ldots, a_{d-1} \in \mathbb{F}_q} \chi(G(a_1, \ldots, a_{d-1})) \right) = 0,$$

where we used Lemma 4.4.1 in the last step.

We can now apply Lemma 4.4.2. This yields the existence of complex numbers $\beta_1, \ldots, \beta_{d-1}$ such that

$$L_s = -\sum_{l=1}^{d-1} \beta_l^s \quad \text{for all } s \geq 1. \tag{4.20}$$

Finally, we evaluate

$$L_s = \sum_P \deg(P)\,\lambda(P^{s/\deg(P)}),$$

where the sum is extended over all monic irreducible polynomials P in $\mathbb{F}_q[x]$ with $\deg(P)$ dividing s. For such a P, let $\gamma \in \mathbb{F}_{q^s}$ be a root of P. Then $P^{s/\deg(P)}$ is the characteristic polynomial of γ over $\mathbb{F}_q$, that is,

$$P(x)^{s/\deg(P)} = (x - \gamma)(x - \gamma^q) \cdots (x - \gamma^{q^{s-1}}),$$

and so

$$\lambda(P^{s/\deg(P)}) = \chi(f(\gamma) + f(\gamma^q) + \cdots + f(\gamma^{q^{s-1}})).$$

The last expression remains the same if γ is replaced by any of its distinct conjugates $\gamma^q, \gamma^{q^2}, \ldots, \gamma^{q^{\deg(P)-1}}$ over $\mathbb{F}_q$. Hence, we can write

$$\deg(P)\,\lambda(P^{s/\deg(P)}) = \sum_{\substack{\gamma \in \mathbb{F}_{q^s} \\ P(\gamma)=0}} \chi\big(f(\gamma) + f(\gamma^q) + \cdots + f(\gamma^{q^{s-1}})\big)$$

and

$$L_s = \sum_P \sum_{\substack{\gamma \in \mathbb{F}_{q^s} \\ P(\gamma)=0}} \chi\big(f(\gamma) + f(\gamma^q) + \cdots + f(\gamma^{q^{s-1}})\big)$$

$$= \sum_{\gamma \in \mathbb{F}_{q^s}} \chi\big(f(\gamma) + f(\gamma^q) + \cdots + f(\gamma^{q^{s-1}})\big).$$

Since f is a polynomial over $\mathbb{F}_q$, we have

$$\chi\big(f(\gamma) + f(\gamma^q) + \cdots + f(\gamma^{q^{s-1}})\big) = \chi\big(f(\gamma) + f(\gamma)^q + \cdots + f(\gamma)^{q^{s-1}}\big)$$

$$= \chi\big(\mathrm{Tr}_s(f(\gamma))\big) = \chi^{(s)}(f(\gamma))$$

for all $\gamma \in \mathbb{F}_{q^s}$, and so

$$L_s = \sum_{\gamma \in \mathbb{F}_{q^s}} \chi^{(s)}(f(\gamma)).$$

In view of (4.20), this completes the proof.   □

We are now ready to prove the desired bound on additive character sums with polynomial arguments. As mentioned earlier in this section, this bound is a consequence of the Hasse-Weil bound in Theorem 4.2.4.

Theorem 4.4.5. Let $f \in \mathbb{F}_q[x]$ be of degree $d \geq 1$ with $\gcd(d, q) = 1$ and let χ be a nontrivial additive character of $\mathbb{F}_q$. Then

$$\left| \sum_{c \in \mathbb{F}_q} \chi(f(c)) \right| \leq (d - 1)q^{1/2}.$$

Proof. First we show that the polynomial $y^q - y - f(x)$ is absolutely irreducible over $\mathbb{F}_q$. Suppose, on the contrary, that we have a nontrivial factorization

$$y^q - y - f(x) = A(x, y)\, B(x, y)$$

over the algebraic closure $\overline{\mathbb{F}_q}$. This means that if k is the degree of $A(x, y)$ in y and m is the degree of $B(x, y)$ in y, then $1 \leq k < q$, $1 \leq m < q$, and $k + m = q$. Let $a(x)$, respectively $b(x)$, be the sum of the terms in $A(x, y)$, respectively $B(x, y)$, that are independent of y. Then $a(x)\, b(x) = -f(x)$, and so

$$
\begin{aligned}
d &= \deg(a) + \deg(b) \\
&\leq k \max\left(\frac{\deg(a)}{k}, \frac{\deg(b)}{m} \right) + m \max\left(\frac{\deg(a)}{k}, \frac{\deg(b)}{m} \right) \\
&= q \max\left(\frac{\deg(a)}{k}, \frac{\deg(b)}{m} \right).
\end{aligned}
$$

Now we carry out the substitution $x \mapsto x^q$, $y \mapsto y^d$. Then

$$y^{dq} - y^d - f(x^q) = A(x^q, y^d)\, B(x^q, y^d).$$

The left-hand side has total degree dq. On the right-hand side, the total degrees satisfy $\deg(A(x^q, y^d)) \geq dk$ and $\deg(B(x^q, y^d)) \geq dm$; hence,

$$dq = \deg(A(x^q, y^d)) + \deg(B(x^q, y^d)) \geq dk + dm = dq,$$

and so $\deg(A(x^q, y^d)) = dk$ and $\deg(B(x^q, y^d)) = dm$. In particular, we obtain $q \deg(a) = \deg(a(x^q)) \leq dk$ and $q \deg(b) = \deg(b(x^q)) \leq dm$, thus

$$d \geq q \max\left(\frac{\deg(a)}{k}, \frac{\deg(b)}{m}\right).$$

Together with the inequality in the opposite direction shown earlier in the proof, this yields

$$\max\left(\frac{\deg(a)}{k}, \frac{\deg(b)}{m}\right) = \frac{d}{q}.$$

But $\gcd(d, q) = 1$, and so the conditions on k and m imply that this identity between rational numbers is impossible. Therefore, $y^q - y - f(x)$ is indeed absolutely irreducible over $\mathbb{F}_q$.

For any integer $s \geq 1$, let N_s be the number of solutions of $y^q - y = f(x)$ in $\mathbb{F}_{q^s} \times \mathbb{F}_{q^s}$, that is,

$$N_s = \sum_{\gamma \in \mathbb{F}_{q^s}} \#\{\alpha \in \mathbb{F}_{q^s} : \alpha^q - \alpha = f(\gamma)\}.$$

We observe that by Theorem 1.4.3(ii), $\alpha^q - \alpha = f(\gamma)$ is solvable in $\mathbb{F}_{q^s}$ if and only if $\mathrm{Tr}_s(f(\gamma)) = 0$. If γ is fixed and $\alpha^q - \alpha = f(\gamma)$ has a solution α, then all solutions are exactly given by $\alpha + u$ with $u \in \mathbb{F}_q$. Therefore,

$$N_s = q \cdot \#\{\gamma \in \mathbb{F}_{q^s} : \mathrm{Tr}_s(f(\gamma)) = 0\}$$

$$= \sum_{\gamma \in \mathbb{F}_{q^s}} \sum_{c \in \mathbb{F}_q} \chi\big(c \, \mathrm{Tr}_s(f(\gamma))\big),$$

where we applied Lemma 4.4.1 in the second step. Note that $\zeta_c(h) = \chi(c\,h)$ for all $h \in \mathbb{F}_q$ defines an additive character of $\mathbb{F}_q$, which is nontrivial if and only if $c \neq 0$. Using the notation in (4.19), we get

$$N_s = \sum_{\gamma \in \mathbb{F}_{q^s}} \sum_{c \in \mathbb{F}_q} \zeta_c^{(s)}(f(\gamma)) = \sum_{c \in \mathbb{F}_q} \sum_{\gamma \in \mathbb{F}_{q^s}} \zeta_c^{(s)}(f(\gamma)),$$

and so

$$N_s - q^s = \sum_{c \in \mathbb{F}_q^*} \sum_{\gamma \in \mathbb{F}_{q^s}} \zeta_c^{(s)}(f(\gamma)).$$

Note that we can assume $d \geq 2$ since the result of the theorem is trivial for $d = 1$. Then by Proposition 4.4.4, for each $c \in \mathbb{F}_q^*$ there exist $\beta_{1,c}, \ldots, \beta_{d-1,c} \in \mathbb{C}$ such that for all $s \geq 1$ we have

$$\sum_{\gamma \in \mathbb{F}_{q^s}} \zeta_c^{(s)}(f(\gamma)) = -\sum_{l=1}^{d-1} \beta_{l,c}^s. \tag{4.21}$$

Therefore,

$$N_s - q^s = -\sum_{c \in \mathbb{F}_q^*} \sum_{l=1}^{d-1} \beta_{l,c}^s \quad \text{for all } s \geq 1.$$

Since $y^q - y - f(x)$ is absolutely irreducible over $\mathbb{F}_q$ by the first part of the proof, the equation $y^q - y = f(x)$ defines a global function field $\mathbb{F}_q(x, y)$ with full constant field $\mathbb{F}_q$. By applying the Hasse-Weil bound in Theorem 4.2.4 to $\mathbb{F}_q(x, y)$ and its constant field extensions, we obtain $N_s - q^s = O(q^{s/2})$ for all $s \geq 1$, with an implied constant which is independent of s in view of [117, Theorem III.6.3(b)]. Thus, the complex power series

$$\sum_{s=1}^{\infty} (N_s - q^s) z^s = \sum_{c \in \mathbb{F}_q^*} \sum_{l=1}^{d-1} \frac{\beta_{l,c} \, z}{\beta_{l,c} \, z - 1}$$

converges absolutely for $|z| < q^{-1/2}$, and so the rational function on the right-hand side cannot have any pole in the disc $|z| < q^{-1/2}$. This means that $|\beta_{l,c}| \leq q^{1/2}$ for all $c \in \mathbb{F}_q^*$ and $1 \leq l \leq d - 1$. Now an application of (4.21) with $c = 1 \in \mathbb{F}_q$ and $s = 1$ completes the proof. $\quad\square$

Remark 4.4.6. The condition $\gcd(d, q) = 1$ in Theorem 4.4.5 can be relaxed in various ways. The following argument yields an easy generalization of Theorem 4.4.5. Let $f \in \mathbb{F}_q[x]$ again be a polynomial of degree $d \geq 1$. Since

$$\chi(f(c) - f(0)) = \chi(f(c))\overline{\chi(f(0))}$$

for all $c \in \mathbb{F}_q$, we can assume that $f(0) = 0$. Then we can write

$$f(x) = \sum_{j \in J} b_j x^j,$$

where the set $J \subseteq \{1, \ldots, d\}$ is such that $b_j \in \mathbb{F}_q^*$ for all $j \in J$. Let p be the characteristic of $\mathbb{F}_q$ and put each $j \in J$ in the form $j = p^{e_j} h_j$ with integers $e_j \geq 0$ and $h_j \geq 1$ satisfying $\gcd(h_j, p) = 1$. Furthermore, we have $\chi = \chi_a$ for some χ_a in (4.14) with $a \in \mathbb{F}_q^*$. For all $j \in J$ and $c \in \mathbb{F}_q$, we obtain

$$\mathrm{Tr}(ab_j c^j) = \mathrm{Tr}\big(ab_j(c^{h_j})^{p^{e_j}}\big) = \mathrm{Tr}\big((u_j c^{h_j})^{p^{e_j}}\big) = \mathrm{Tr}(u_j c^{h_j})$$

with $u_j \in \mathbb{F}_q^*$ independent of c. From (4.14) we get then

$$\chi(b_j c^j) = \chi_1(ab_j c^j) = \chi_1(u_j c^{h_j}) = \chi(a^{-1}u_j c^{h_j})$$

for all $j \in J$ and $c \in \mathbb{F}_q$, and so

$$\chi(f(c)) = \prod_{j \in J} \chi(b_j c^j) = \prod_{j \in J} \chi(a^{-1}u_j c^{h_j}) = \chi(g(c))$$

for all $c \in \mathbb{F}_q$, where

$$g(x) = \sum_{j \in J} a^{-1}u_j x^{h_j}.$$

Now assume that among the numbers h_j, $j \in J$, there is a unique largest one. Then $\deg(g) \geq 1$ and $\gcd(\deg(g), q) = 1$, so that Theorem 4.4.5 can be applied to the character sum $\sum_{c \in \mathbb{F}_q} \chi(g(c))$. This yields

$$\left| \sum_{c \in \mathbb{F}_q} \chi(f(c)) \right| = \left| \sum_{c \in \mathbb{F}_q} \chi(g(c)) \right| \leq (\deg(g) - 1)q^{1/2} \leq (d - 1)q^{1/2},$$

and so a generalization of Theorem 4.4.5.

We now consider the analog of the character sum in Theorem 4.4.5 for multiplicative characters of $\mathbb{F}_q$. We start with some preliminaries. It is clear from the explicit formula (4.15) that the multiplicative characters of $\mathbb{F}_q$ form a cyclic group of order $q - 1$ under multiplication. Thus, we can speak of the *order* of a multiplicative character as the order in this group. According to a standard convention, for a nontrivial multiplicative character ψ of $\mathbb{F}_q$ we define $\psi(0) = 0$ and for the trivial multiplicative character ψ_0 of $\mathbb{F}_q$ we define $\psi_0(0) = 1$. Note that then $\psi(bc) = \psi(b)\psi(c)$ for all $b, c \in \mathbb{F}_q$ and any multiplicative character ψ of $\mathbb{F}_q$.

A given multiplicative character ψ of $\mathbb{F}_q$ can be lifted canonically to a multiplicative character $\psi^{(s)}$ of an extension field $\mathbb{F}_{q^s}$ of $\mathbb{F}_q$ by using the norm map $\mathrm{Nm}_s : \mathbb{F}_{q^s} \to \mathbb{F}_q$ from $\mathbb{F}_{q^s}$ to $\mathbb{F}_q$. Recall from Section 1.4 that

$$\mathrm{Nm}_s(\gamma) = \prod_{i=0}^{s-1} \gamma^{q^i} = \gamma^{(q^s-1)/(q-1)} \quad \text{for all } \gamma \in \mathbb{F}_{q^s}.$$

We define

$$\psi^{(s)}(\gamma) = \psi(\mathrm{Nm}_s(\gamma)) \quad \text{for all } \gamma \in \mathbb{F}_{q^s}. \tag{4.22}$$

Lemma 4.4.7. Let $c \in \mathbb{F}_q$ and let m be a positive integer dividing $q - 1$. Then the number of solutions of $y^m = c$ in $\mathbb{F}_q$ is equal to

$$\sum_{\psi \in X_m} \psi(c),$$

where X_m is the set of all multiplicative characters of $\mathbb{F}_q$ of order dividing m.

Proof. Let $\eta = \psi_{(q-1)/m}$ in the notation of (4.15). Then η is a multiplicative character of $\mathbb{F}_q$ of order m and $X_m = \{\eta^j : j = 0, 1, \ldots, m - 1\}$. Write $S(c)$ for the sum in the lemma and $N(c)$ for the number of solutions of $y^m = c$ in $\mathbb{F}_q$. If $c = 0$, then $S(c) = 1 = N(c)$. If c is a nonzero mth power, then $\psi(c) = 1$ for all $\psi \in X_m$ and $S(c) = m = N(c)$. If c is not an mth power, then $\eta(c) \neq 1$ and

$$S(c) = \sum_{j=0}^{m-1} \eta^j(c) = 0 = N(c).$$

Thus, in all cases we have $S(c) = N(c)$. $\qquad\square$

Corollary 4.4.8. Let $s \geq 1$ be an integer, let $\gamma \in \mathbb{F}_{q^s}$, and let m be a positive integer dividing $q - 1$. Then the number of solutions of $y^m = \gamma$ in $\mathbb{F}_{q^s}$ is equal to

$$\sum_{\psi \in X_m} \psi^{(s)}(\gamma),$$

where X_m is the set of all multiplicative characters of $\mathbb{F}_q$ of order dividing m.

Proof. By Lemma 4.4.7 and its proof, the number of solutions of $y^m = \gamma$ in $\mathbb{F}_{q^s}$ is equal to

$$\sum_{\psi \in X_m^{(s)}} \psi(\gamma),$$

where $X_m^{(s)} = \{\kappa^j : j = 0, 1, \ldots, m - 1\}$ and $\kappa = \psi_{(q^s-1)/m}$ in the notation of (4.15), with q replaced by q^s and with the primitive element ρ of $\mathbb{F}_{q^s}$, that is,

$$\kappa(\rho^k) = e^{2\pi i k/m} \quad \text{for } k = 0, 1, \ldots, q^s - 2.$$

Then $r = \rho^{(q^s-1)/(q-1)}$ is a primitive element of $\mathbb{F}_q$. Let η be the multiplicative character of $\mathbb{F}_q$ given by $\eta = \psi_{(q-1)/m}$ in (4.15) with this choice for r. Then for $k = 0, 1, \ldots, q^s - 2$ we have by (4.22),

$$\eta^{(s)}(\rho^k) = \eta\big(\mathrm{Nm}_s(\rho^k)\big) = \eta(\rho^{k(q^s-1)/(q-1)})$$
$$= \eta(r^k) = e^{2\pi i k/m} = \kappa(\rho^k),$$

and so $\eta^{(s)} = \kappa$. This shows that $X_m^{(s)} = \{\psi^{(s)} : \psi \in X_m\}$. $\square$

Let now ψ be a nontrivial multiplicative character of $\mathbb{F}_q$ and let $f \in \mathbb{F}_q[x]$ be of positive degree. We consider the character sum

$$\sum_{c \in \mathbb{F}_q} \psi(f(c)).$$

It is clear that we can assume without loss of generality that f is monic, and this will be done in the sequel. If m is the order of ψ and f is an mth power of a polynomial, then we get a trivial situation. Thus, we will assume that

f is not an mth power of a polynomial. The following result is an analog of Proposition 4.4.4.

Proposition 4.4.9. Let ψ be a multiplicative character of $\mathbb{F}_q$ of order $m > 1$ and let $f \in \mathbb{F}_q[x]$ be a monic polynomial of positive degree, which is not an mth power of a polynomial. Let k be the number of distinct roots of f in its splitting field over $\mathbb{F}_q$ and suppose that $k \geq 2$. Then there exist complex numbers $\beta_1, \ldots, \beta_{k-1}$, depending only on f and ψ, such that for any integer $s \geq 1$ we have

$$\sum_{\gamma \in \mathbb{F}_{q^s}} \psi^{(s)}(f(\gamma)) = -\sum_{l=1}^{k-1} \beta_l^s.$$

Proof. We define a complex-valued function λ on M_q as follows. We put $\lambda(1) = 1$. If $g \in M_q^{(n)}$ for some $n \geq 1$, then let $g(x) = (x - \alpha_1) \cdots (x - \alpha_n)$ be the factorization of g in its splitting field over $\mathbb{F}_q$. Note that $f(\alpha_1) \cdots f(\alpha_n) \in \mathbb{F}$, since this element is the resultant $R(g, f)$ of g and f. We put

$$\lambda(g) = \psi(f(\alpha_1) \cdots f(\alpha_n)).$$

Then (4.16) is satisfied. Let

$$f = f_1^{e_1} \cdots f_r^{e_r} \tag{4.23}$$

be the canonical factorization of f in $\mathbb{F}_q[x]$, where $f_1, \ldots, f_r$ are distinct monic irreducible polynomials in $\mathbb{F}_q[x]$. Then we have

$$\lambda(g) = \psi^{e_1}(R(g, f_1)) \cdots \psi^{e_r}(R(g, f_r)).$$

For $1 \leq i \leq r$, let $d_i = \deg(f_i)$ and let μ_i be a root of f_i in the finite field E_i of order q^{d_i}. Then

$$R(g, f_i) = (-1)^{n d_i} R(f_i, g) = (-1)^{n d_i} \prod_{j=0}^{d_i - 1} g(\mu_i^{q^j})$$

$$= (-1)^{n d_i} \prod_{j=0}^{d_i - 1} g(\mu_i)^{q^j} = (-1)^{n d_i} \mathrm{Nm}_{d_i}(g(\mu_i)),$$

and so

$$\lambda(g) = \psi((-1)^{nd}) \prod_{i=1}^{r} \psi^{e_i}(\mathrm{Nm}_{d_i}(g(\mu_i)))$$

with $d = \deg(f)$. Writing τ_i for the multiplicative character obtained by lifting ψ^{e_i} to E_i and $\varepsilon_n = \psi((-1)^{nd})$, we thus have

$$\lambda(g) = \varepsilon_n \prod_{i=1}^{r} \tau_i(g(\mu_i)). \tag{4.24}$$

Since, by hypothesis, f is not an mth power, at least one of the e_i in (4.23) is not a multiple of m; hence, at least one of the ψ^{e_i}, and so at least one of the τ_i, is nontrivial.

Next we verify (4.17) for all $n \geq k$. We note that $k = d_1 + \cdots + d_r$. Let the map $\sigma : M_q^{(n)} \rightarrow E_1 \times \cdots \times E_r$ be defined by

$$\sigma(g) = (g(\mu_1), \ldots, g(\mu_r)) \quad \text{for } g \in M_q^{(n)}.$$

Let $(v_1, \ldots, v_r) \in E_1 \times \cdots \times E_r$ be given. Each v_i, $1 \leq i \leq r$, can be represented in the form $v_i = h_i(\mu_i)$ with $h_i \in \mathbb{F}_q[x]$. Then $\sigma(g) = (v_1, \ldots, v_r)$ if and only if g is a solution of the system of congruences

$$g \equiv h_i \pmod{f_i} \quad \text{for } 1 \leq i \leq r.$$

By the Chinese remainder theorem, this system of congruences has a unique solution $G \in \mathbb{F}_q[x]$ with $\deg(G) < d_1 + \cdots + d_r = k$. Then all solutions $g \in M_q^{(n)}$ of the system are given by $g = F f_1 \cdots f_r + G$, where F is an arbitrary monic polynomial over $\mathbb{F}_q$ of degree $n-k$. Since there are exactly q^{n-k} choices for F, there are exactly q^{n-k} polynomials $g \in M_q^{(n)}$ with $\sigma(g) = (g(\mu_1), \ldots, g(\mu_r)) = (v_1, \ldots, v_r)$. Using this fact and (4.24), we get

$$\sum_{g \in M_q^{(n)}} \lambda(g) = \varepsilon_n q^{n-k} \sum_{v_1 \in E_1} \cdots \sum_{v_r \in E_r} \tau_1(v_1) \cdots \tau_r(v_r)$$

$$= \varepsilon_n q^{n-k} \left(\sum_{v_1 \in E_1} \tau_1(v_1) \right) \cdots \left(\sum_{v_r \in E_r} \tau_r(v_r) \right) = 0,$$

where the last step follows from Lemma 4.4.1 since at least one of the τ_i is nontrivial, as we have noted before.

We can now apply Lemma 4.4.2. This yields the existence of complex numbers $\beta_1, \ldots, \beta_{k-1}$ such that

$$L_s = -\sum_{l=1}^{k-1} \beta_l^s \quad \text{for all } s \geq 1. \tag{4.25}$$

Finally, we evaluate

$$L_s = \sum_P \deg(P)\, \lambda(P^{s/\deg(P)}),$$

where the sum is extended over all monic irreducible polynomials P in $\mathbb{F}_q[x]$ with $\deg(P)$ dividing s. As in the proof of Proposition 4.4.4, we see that

$$L_s = \sum_{\gamma \in \mathbb{F}_{q^s}} \psi\big(f(\gamma)f(\gamma^q)\cdots f(\gamma^{q^{s-1}})\big).$$

Since f is a polynomial over $\mathbb{F}_q$, we have

$$\psi\big(f(\gamma)f(\gamma^q)\cdots f(\gamma^{q^{s-1}})\big) = \psi\big(f(\gamma)f(\gamma)^q\cdots f(\gamma)^{q^{s-1}}\big)$$
$$= \psi\big(\mathrm{Nm}_s(f(\gamma))\big) = \psi^{(s)}(f(\gamma))$$

for all $\gamma \in \mathbb{F}_{q^s}$, and so

$$L_s = \sum_{\gamma \in \mathbb{F}_{q^s}} \psi^{(s)}(f(\gamma)).$$

In view of (4.25), this completes the proof. □

We are now ready to prove the desired bound on multiplicative character sums with polynomial arguments. This bound is again a consequence of the Hasse-Weil bound in Theorem 4.2.4.

Theorem 4.4.10. Let ψ be a multiplicative character of $\mathbb{F}_q$ of order $m > 1$ and let $f \in \mathbb{F}_q[x]$ be a monic polynomial of positive degree, which is not an mth power of a polynomial. Let k be the number of distinct roots of f in its splitting field over $\mathbb{F}_q$. Then

$$\left| \sum_{c \in \mathbb{F}_q} \psi(f(c)) \right| \leq (k-1)q^{1/2}.$$

Proof. Let the canonical factorization of f in $\mathbb{F}_q[x]$ be as in (4.23). Put $D = \gcd(m, e_1, \ldots, e_r)$ and note that $D < m$ since f is not an mth power. Moreover, we can write $f = h^D$ with

$$h = f_1^{e_1/D} \cdots f_r^{e_r/D}.$$

Then we have

$$\sum_{c \in \mathbb{F}_q} \psi(f(c)) = \sum_{c \in \mathbb{F}_q} \psi^D(h(c)).$$

Note that ψ^D has order m/D and

$$\gcd\left(\frac{m}{D}, \frac{e_1}{D}, \ldots, \frac{e_r}{D}\right) = 1.$$

Thus, we can reduce the general case to the case where $\gcd(m, e_1, \ldots, e_r) = 1$.

Under the condition $\gcd(m, e_1, \ldots, e_r) = 1$, we now show that the polynomial $y^m - f(x)$ is absolutely irreducible over $\mathbb{F}_q$. Put $K = \overline{\mathbb{F}_q}(x)$ and assume that $y^m - f(x)$ is reducible in $K[y]$. Let ζ be a primitive mth root of unity in $\mathbb{F}_q$. Then over $\overline{K}$ we have a factorization of the form

$$y^m - f(x) = \prod_{j=0}^{m-1} (y - \zeta^j Y),$$

Since $y^m - f(x)$ is reducible in $K[y]$, there exists a product of the form

$$(y - \zeta^{j_1} Y) \cdots (y - \zeta^{j_t} Y),$$

which belongs to $K[y]$, where $0 \leq j_1 < j_2 < \cdots < j_t < m$ and $1 \leq t < m$. The constant term

$$(-1)^t \zeta^{j_1 + \cdots + j_t} Y^t$$

of this product belongs to K, and so $Y^t \in K$. Let u be the least positive integer for which $Y^u \in K$. Then $u \le t < m$ and u divides m since $Y^m = f(x) \in K$. For $w = Y^u$ we have $w^{m/u} = Y^m = f$. From the uniqueness of factorization in $\overline{\mathbb{F}_q}[x]$ it follows that m/u divides e_i for $1 \le i \le r$, and so

$$\gcd(m, e_1, \ldots, e_r) \ge \frac{m}{u} > 1,$$

a contradiction. Thus, $y^m - f(x)$ is indeed absolutely irreducible over $\mathbb{F}_q$.

For any integer $s \ge 1$, let N_s be the number of solutions of $y^m = f(x)$ in $\mathbb{F}_{q^s} \times \mathbb{F}_{q^s}$, that is,

$$N_s = \sum_{\gamma \in \mathbb{F}_{q^s}} \#\{\alpha \in \mathbb{F}_{q^s} : \alpha^m = f(\gamma)\}.$$

Then by Corollary 4.4.8 we have

$$N_s = \sum_{\gamma \in \mathbb{F}_{q^s}} \sum_{\psi \in X_m} \psi^{(s)}(f(\gamma)) = \sum_{\psi \in X_m} \sum_{\gamma \in \mathbb{F}_{q^s}} \psi^{(s)}(f(\gamma)),$$

and so

$$N_s - q^s = \sum_{\psi \in X_m^*} \sum_{\gamma \in \mathbb{F}_{q^s}} \psi^{(s)}(f(\gamma)),$$

where $X_m^* := X_m \setminus \{\psi_0\}$ is the set of nontrivial characters in X_m. The result of the theorem is trivial in the case $k = 1$, and so we can assume $k \ge 2$. Then Proposition 4.4.9 shows that for each $\psi \in X_m^*$ there exist $\beta_{1,\psi}, \ldots, \beta_{k-1,\psi} \in \mathbb{C}$ such that for all $s \ge 1$ we have

$$\sum_{\gamma \in \mathbb{F}_{q^s}} \psi^{(s)}(f(\gamma)) = -\sum_{l=1}^{k-1} \beta_{l,\psi}^s.$$

Therefore,

$$N_s - q^s = -\sum_{\psi \in X_m^*} \sum_{l=1}^{k-1} \beta_{l,\psi}^s \quad \text{for all } s \ge 1.$$

Since $y^m - f(x)$ is absolutely irreducible over $\mathbb{F}_q$ by an earlier part of the proof, the equation $y^m = f(x)$ defines a global function field $\mathbb{F}_q(x, y)$ with full constant field $\mathbb{F}_q$. By applying the Hasse-Weil bound in Theorem 4.2.4 to $\mathbb{F}_q(x, y)$ and its constant field extensions, we obtain $N_s - q^s = O(q^{s/2})$ for all $s \geq 1$, with an implied constant, which is independent of s in view of [117, Theorem III.6.3(b)]. The proof is now completed as in the proof of Theorem 4.4.5.   □

Results on related character sums over finite fields can be found in the books of Li [68], Lidl and Niederreiter [72], and Schmidt [109].

5   Applications to Coding Theory

A code is a scheme for detecting and correcting transmission errors that can occur in noisy communication channels. A code operates by adding redundant information to messages. One of the main aims of coding theory is the construction of efficient codes, that is, of codes that achieve a pre-scribed error-correction capability with a minimum amount of redundancy. Algebraic curves over finite fields, or equivalently global function fields, play an important role in coding theory, as we will demonstrate in this chapter.

We start this chapter by recalling some basic facts from coding theory in Section 5.1. A core part of the chapter is Section 5.2 in which algebraic-geometry codes are introduced. This powerful family of codes is constructed from algebraic curves over finite fields or equivalently from global function fields. Several major results in the asymptotic theory of codes, including some quite recent ones, are proved in Section 5.3. These results provide evidence that algebraic-geometry codes are very competitive when it comes to the construction of efficient codes of large length. Various extensions of the construction of algebraic-geometry codes are presented in Sections 5.4 and 5.5. Applications of character sums to coding theory are discussed in Section 5.6. Finally, Section 5.7 builds on the analogy between linear codes and digital nets and shows constructions of digital nets that are similar to those of algebraic-geometry codes.

We refer to the books of Ling and Xing [73], van Lint [74], and MacWilliams and Sloane [75] for detailed background on coding theory.

5.1 Background on Codes

As the signal alphabet for our codes we always use $\mathbb{F}_q$, the finite field of order q. For a positive integer n, let $\mathbb{F}_q^n$ denote the set (and also the additive group) of n-tuples of elements of $\mathbb{F}_q$. Mathematically, a *code* C over $\mathbb{F}_q$ is simply a nonempty subset of $\mathbb{F}_q^n$. The number $n = n(C)$ is called the *length* of the code C. An element of C is called a *codeword* of C.

We write $\log_q$ for the logarithm to the base q. Then the *information rate* $R(C)$ of the code C is defined to be

$$R(C) = \frac{\log_q |C|}{n(C)}.$$

It is one aspect of measuring the efficiency of a code. A higher information rate means less redundancy in the code.

Example 5.1.1. Let $q = 2, n = 3$, and $C = \{(0, 0, 0), (1, 1, 1)\} \subseteq \mathbb{F}_2^3$. A binary message, that is, a finite string of bits, is encoded bitwise by the map $\{0, 1\} \to C$ given by the assignments $0 \mapsto (0, 0, 0)$ and $1 \mapsto (1, 1, 1)$. The information rate of C is $R(C) = \frac{1}{3}$, which can be seen to mean that each bit of each codeword of C carries $\frac{1}{3}$ bit of actual information. It is easily checked that C allows us to correct one error in a block of 3 bits. For instance, if we want to send the message 1, then it is encoded into the codeword $(1, 1, 1)$ before transmission. Suppose that at the other end of the communication channel the triple $(1, 0, 1)$ is received. If we assume that at most one error has occurred during the transmission, then the sent codeword could not have been $(0, 0, 0)$. Thus, the recipient can uniquely recover the codeword $(1, 1, 1)$, which corresponds to the message 1.

To determine the error-correction capability of an arbitrary code, we introduce several simple concepts. First of all, for $\mathbf{x} \in \mathbb{F}_q^n$ the *(Hamming) weight* $w(\mathbf{x})$ is the number of nonzero coordinates of $\mathbf{x}$. For $\mathbf{x}, \mathbf{y} \in \mathbb{F}_q^n$ the *(Hamming) distance* $d(\mathbf{x}, \mathbf{y})$ is given by

$$d(\mathbf{x}, \mathbf{y}) = w(\mathbf{x} - \mathbf{y}).$$

In words, $d(\mathbf{x}, \mathbf{y})$ is the number of coordinates in which $\mathbf{x}$ and $\mathbf{y}$ differ. Note that d is a metric on $\mathbb{F}_q^n$. For a code $C \subseteq \mathbb{F}_q^n$ with $|C| \geq 2$, we define its *minimum distance*

$$d(C) = \min \{d(\mathbf{x}, \mathbf{y}) : \mathbf{x}, \mathbf{y} \in C, \ \mathbf{x} \neq \mathbf{y}\}.$$

It is a basic fact that C can correct up to $\lfloor (d(C) - 1)/2 \rfloor$ transmission errors in each received n-tuple over $\mathbb{F}_q$. For instance, the code C in Example 5.1.1 has minimum distance $d(C) = 3$, and so it can correct one error in each received triple as mentioned in Example 5.1.1.

The set $\mathbb{F}_q^n$ is endowed with the structure of a vector space over $\mathbb{F}_q$. Particularly nice examples of codes $C \subseteq \mathbb{F}_q^n$ are those that form a nonzero $\mathbb{F}_q$-linear subspace of the vector space $\mathbb{F}_q^n$. Such a code C is called a *linear code* over $\mathbb{F}_q$. If C is a linear code over $\mathbb{F}_q$, then its dimension as a vector space over $\mathbb{F}_q$ is called the *dimension* of the code C. If the length of C is n and its dimension is k, then we say that C is a linear $[n, k]$ code over $\mathbb{F}_q$. Note that by definition we have $1 \leq k \leq n$. Moreover, for the information rate $R(C)$ of C we have then the simple expression

$$R(C) = \frac{k}{n}.$$

If we know in addition that C has minimum distance $d = d(C)$, then we say that C is a linear $[n, k, d]$ code over $\mathbb{F}_q$. It is an easy observation that for a linear code C we have

$$d(C) = \min_{\mathbf{x} \in C \setminus \{\mathbf{0}\}} w(\mathbf{x}). \tag{5.1}$$

A linear code over $\mathbb{F}_q$ can also be viewed as the null space of a suitable matrix over $\mathbb{F}_q$. If C is a linear $[n, k]$ code over $\mathbb{F}_q$, then any matrix H over $\mathbb{F}_q$ with n columns, which has C as its null space, that is,

$$C = \left\{ \mathbf{x} \in \mathbb{F}_q^n : H\mathbf{x}^{\mathrm{T}} = \mathbf{0} \right\},$$

is called a *parity-check matrix* of C. Note that H must have rank $n - k$. Often, but not always, it is assumed that H is an $(n - k) \times n$ matrix. The minimum distance of C can be read off from any parity-check matrix of C, according to the following result.

Proposition 5.1.2. Let H be any parity-check matrix of the linear code C over $\mathbb{F}_q$. Then C has minimum distance d if and only if any $d - 1$ columns of H are linearly independent and some d columns of H are linearly dependent.

Proof. There exists a $\mathbf{c} \in C$ of some weight $w \geq 1$ if and only if $H\mathbf{c}^{\mathrm{T}} = \mathbf{0}$ for some $\mathbf{c} \in \mathbb{F}_q^n$ of weight w, and this is in turn equivalent to some w columns of H being linearly dependent. In view of (5.1), this yields the result of the proposition. $\qquad\square$

There is a simple inequality linking the three parameters n, k, and d of a linear code.

Proposition 5.1.3 (Singleton Bound). For any linear $[n, k, d]$ code over $\mathbb{F}_q$, we have

$$d \le n - k + 1.$$

Proof. Let H be a parity-check matrix of a given linear $[n, k, d]$ code over $\mathbb{F}_q$. Then $n - k$ is the rank of H, and so any $n - k + 1$ columns of H are linearly dependent. Thus $d \le n - k + 1$ by Proposition 5.1.2. $\square$

Remark 5.1.4. A linear $[n, k, d]$ code over $\mathbb{F}_q$ with $d = n - k + 1$ is called an *MDS code*. Here MDS stands for "maximum distance separable." The linear $[3, 1, 3]$ code over $\mathbb{F}_2$ in Example 5.1.1 provides a simple example of an MDS code.

Example 5.1.5. Let k be an integer with $1 \le k \le q$. The set $\mathcal{P}_k$ of polynomials over $\mathbb{F}_q$ of degree at most $k - 1$ forms a vector space over $\mathbb{F}_q$ of dimension k. Let $b_1, \ldots, b_q$ be all elements of $\mathbb{F}_q$. Define the linear code C over $\mathbb{F}_q$ as the image of the $\mathbb{F}_q$-linear map

$$\psi : f \in \mathcal{P}_k \mapsto (f(b_1), \ldots, f(b_q)) \in \mathbb{F}_q^q.$$

For a nonzero $f \in \mathcal{P}_k$, the weight w of $\psi(f)$ is given by $w = q - z$, where z is the number of zeros of f in $\mathbb{F}_q$. Now $z \le k - 1$, and so $w \ge q - k + 1 > 0$. This shows that $d(C) \ge q - k + 1$ and also that ψ is injective. Thus, C has dimension k and we get $d(C) = q - k + 1$ by Proposition 5.1.3. Altogether, C is a linear $[q, k, q - k + 1]$ code over $\mathbb{F}_q$, called an (extended) *Reed-Solomon code* over $\mathbb{F}_q$. Clearly, C is an MDS code.

The procedure in the proof of the following result shows how, in principle, many good linear codes can be obtained. However, it should be noted that this procedure is usually not efficient.

Proposition 5.1.6 (Gilbert-Varshamov Bound). Let $n, k,$ and d be integers with $1 \le k < n, 2 \le d \le n$, and

$$\sum_{i=0}^{d-2} \binom{n-1}{i} (q-1)^i < q^{n-k}. \tag{5.2}$$

Then there exists a linear $[n, k]$ code over $\mathbb{F}_q$ with minimum distance at least d.

Proof. We have

$$q^{d-2} = \sum_{i=0}^{d-2} \binom{d-2}{i}(q-1)^i \leq \sum_{i=0}^{d-2}\binom{n-1}{i}(q-1)^i < q^{n-k}$$

by (5.2), and so $d - 1 \leq n - k$. We proceed by constructing a suitable $(n-k) \times n$ parity-check matrix H over $\mathbb{F}_q$ columnwise. We choose the first $d - 1$ columns of H as linearly independent vectors from $\mathbb{F}_q^{n-k}$ (this is possible since $d - 1 \leq n - k$). Now suppose that $j - 1$ columns of H (with $d \leq j \leq n$) have already been constructed so that any $d - 1$ of them are linearly independent. There are at most

$$\sum_{i=0}^{d-2} \binom{j-1}{i}(q-1)^i \leq \sum_{i=0}^{d-2}\binom{n-1}{i}(q-1)^i$$

vectors obtained by linear combinations of $d - 2$ or fewer of these $j - 1$ columns. Since (5.2) holds, it is possible to choose a jth column that is linearly independent of any $d - 2$ of the first $j - 1$ columns of H. The null space of H is a linear code over $\mathbb{F}_q$ of length n, of minimum distance at least d by Proposition 5.1.2, and of dimension at least k. By passing to a k-dimensional subspace, we get a linear code of the desired type. $\qquad\square$

5.2 Algebraic-Geometry Codes

A powerful family of linear codes was constructed by Goppa [43–45] using algebraic curves over finite fields and differentials. Nowadays, the standard way of introducing these codes employs global function fields and Riemann-Roch spaces.

Let $F/\mathbb{F}_q$ be a global function field over $\mathbb{F}_q$ of genus g. We assume that $N(F) \geq 1$, where $N(F)$ denotes, as usual, the number of rational places of F. We will construct a linear code over $\mathbb{F}_q$ of length $n \leq N(F)$. Choose n distinct rational places $P_1, \ldots, P_n$ of F and a divisor G of F with

$\text{supp}(G) \cap \{P_1, \ldots, P_n\} = \varnothing$. Consider the Riemann-Roch space

$$\mathcal{L}(G) = \{f \in F^* : \text{div}(f) + G \geq 0\} \cup \{0\},$$

where $\text{div}(f)$ is the principal divisor of $f \in F^*$. Note that $v_{P_i}(f) \geq 0$ for $1 \leq i \leq n$ and all $f \in \mathcal{L}(G)$. Therefore, the residue class $f(P_i)$ of $f \in \mathcal{O}_{P_i}$, that is, the image of f under the residue class map of the place P_i, is defined (see Definition 1.5.10). Since P_i is a rational place, $f(P_i)$ can be identified with an element of $\mathbb{F}_q$.

Definition 5.2.1. The *algebraic-geometry code* (or *AG code*) $C(P_1, \ldots, P_n; G)$ is defined as the image of the $\mathbb{F}_q$-linear map $\psi : \mathcal{L}(G) \to \mathbb{F}_q^n$ given by

$$\psi(f) = (f(P_1), \ldots, f(P_n)) \quad \text{for all } f \in \mathcal{L}(G). \tag{5.3}$$

The standard result on the parameters of AG codes is the following one.

Theorem 5.2.2. Let $F/\mathbb{F}_q$ be a global function field of genus g with $N(F) \geq g + 1$. For $g < n \leq N(F)$, choose n distinct rational places $P_1, \ldots, P_n$ of F and a divisor G of F with $g \leq \deg(G) < n$ and $\text{supp}(G) \cap \{P_1, \ldots, P_n\} = \varnothing$. Then $C(P_1, \ldots, P_n; G)$ is a linear $[n, k, d]$ code over $\mathbb{F}_q$ with

$$k = \ell(G) \geq \deg(G) + 1 - g, \qquad d \geq n - \deg(G).$$

Moreover, $k = \deg(G) + 1 - g$ if $\deg(G) \geq 2g - 1$.

Proof. For any nonzero $f \in \mathcal{L}(G)$, the weight w of $\psi(f)$ is given by $w = n - z$, where z is the number of zeros of f in the set $\{P_1, \ldots, P_n\}$. If $P_{i_1}, \ldots, P_{i_z}$ are these distinct zeros, then

$$f \in \mathcal{L}(G - P_{i_1} - \cdots - P_{i_z}).$$

Since $f \neq 0$, this implies by Corollary 3.4.4 that

$$0 \leq \deg(G - P_{i_1} - \cdots - P_{i_z}) = \deg(G) - z,$$

and so $w \geq n - \deg(G) > 0$. This shows not only the desired lower bound on d, but also that ψ is injective. Hence, $k = \ell(G)$, and the rest follows from the Riemann-Roch theorem (see Theorem 3.6.14). $\qquad \square$

Remark 5.2.3. Theorem 5.2.2 implies that $k + d \geq n + 1 - g$. This should be compared with the Singleton bound $k + d \leq n + 1$ in Proposition 5.1.3. Thus, the genus of F controls, in a sense, the deviation of $k + d$ from the Singleton bound. If $g = 0$, that is, if F is the rational function field over $\mathbb{F}_q$, then any AG code in Theorem 5.2.2 is an MDS code (compare with Remark 5.1.4).

Example 5.2.4. Let $F = \mathbb{F}_q(x)$ be the rational function field over $\mathbb{F}_q$. Let $\mathbb{F}_q = \{b_1, \ldots, b_q\}$ and for $1 \leq i \leq q$ let P_i be the rational place $x - b_i$ of $\mathbb{F}_q(x)$. Put $G = (k - 1)P_\infty$, where k is an integer with $1 \leq k \leq q$ and P_∞ denotes the infinite place of $\mathbb{F}_q(x)$. Then the AG code $C(P_1, \ldots, P_q; G)$ is the Reed-Solomon code in Example 5.1.5. The argument in the proof of Theorem 5.2.2 can be viewed as a generalization of the argument in Example 5.1.5.

Example 5.2.5. Let F be the Hermitian function field over $\mathbb{F}_{q^2}$, that is, $F = \mathbb{F}_{q^2}(x, y)$ with $y^q + y = x^{q+1}$. Then F has genus $g = (q^2 - q)/2$ and $N(F) = q^3 + 1$ (see [117, Lemma VI.4.4]). Let Q be the rational place of F lying over the infinite place of $\mathbb{F}_{q^2}(x)$ and let $P_1, \ldots, P_n$ with $n = q^3$ be the remaining rational places of F. Put $G = mQ$ with an integer m satisfying $q^2 - q - 1 \leq m < q^3$. Then by Theorem 5.2.2, the AG code $C(P_1, \ldots, P_n; G)$ is a linear $[n, k, d]$ code over $\mathbb{F}_{q^2}$ with $k = m + 1 - (q^2 - q)/2$ and $d \geq q^3 - m$. Such a code is called a *Hermitian code*.

Example 5.2.6. Take $q = 2$ and $m = 4$ in Example 5.2.5, so that $G = 4Q$. Then the Hermitian code $C(P_1, \ldots, P_8; G)$ in Example 5.2.5 is a linear $[8, 4, d]$ code over $\mathbb{F}_4$ with $d \geq 4$. We can show directly that actually $d = 4$. Note first that $f(x) = x(x + 1) \in \mathbb{F}_4[x]$ satisfies $f \in \mathcal{L}(G)$. Furthermore, f has exactly four zeros in F, namely the rational places of F lying over x or $x + 1$. Consequently, the codeword $\psi(f)$ in (5.3) has weight 4, and so $d = 4$. The code $C(P_1, \ldots, P_8; G)$ is optimal in the sense that there is no linear $[8, 4]$ code over $\mathbb{F}_4$ with minimum distance ≥ 5 (see Brouwer [11] and Grassl [46]).

The condition on the support of G in Theorem 5.2.2 is a conventional one in the area. However, as the following result shows, we can easily get rid of this condition.

Theorem 5.2.7. Let $F/\mathbb{F}_q$ be a global function field with $N(F) \geq 1$. For an integer n with $1 \leq n \leq N(F)$, let G be a divisor of F with $\deg(G) < n$ and $\ell(G) \geq 1$. Then there exists an AG code, which is a linear $[n, k, d]$ code over $\mathbb{F}_q$ with

$$k = \ell(G), \qquad d \geq n - \deg(G).$$

Proof. Let $P_1, \ldots, P_n$ be n distinct rational places of F. By the approximation theorem (see Theorem 1.5.18), we can choose $u \in F$ such that $v_{P_i}(u) = v_{P_i}(G)$ for $1 \leq i \leq n$. Put $G' = G - \operatorname{div}(u)$ and note that the divisor G' satisfies $\operatorname{supp}(G') \cap \{P_1, \ldots, P_n\} = \varnothing$. Now consider the AG code $C(P_1, \ldots, P_n; G')$. As in the proof of Theorem 5.2.2, we see that this is a linear $[n, k, d]$ code over $\mathbb{F}_q$ with $k = \ell(G')$ and $d \geq n - \deg(G')$. Since G' is equivalent to G, we have $\ell(G') = \ell(G)$ and $\deg(G') = \deg(G)$, and so the proof is complete. $\square$

Example 5.2.8. Let F be the rational function field over $\mathbb{F}_q$. Then we can choose $n = q + 1$ in Theorem 5.2.7. For an integer k with $1 \leq k \leq q + 1$, let G be a divisor of F with $\deg(G) = k - 1$. Then Theorem 5.2.7 shows the existence of a linear $[q + 1, k, d]$ code over $\mathbb{F}_q$ with $d \geq q - k + 2$. On the other hand, the Singleton bound in Proposition 5.1.3 yields $d \leq q - k + 2$. Thus, $d = q - k + 2$ and this code is an MDS code (see Remark 5.1.4).

Example 5.2.9. Let F be the Hermitian function field over $\mathbb{F}_4$. Since $N(F) = 9$, we can choose $n = 9$ in Theorem 5.2.7. Let G be a divisor of F with $\deg(G) = 4$. Using $g(F) = 1$, we get $\ell(G) = 4$. Then Theorem 5.2.7 shows the existence of a linear $[9, 4, d]$ code over $\mathbb{F}_4$ with $d \geq 5$. But it is known that there is no linear $[9, 4]$ code over $\mathbb{F}_4$ with minimum distance ≥ 6 (see Brouwer [11] and Grassl [46]), and so $d = 5$.

The usefulness of AG codes is enhanced by the fact that there exist efficient algorithms for solving the decoding problem for this family of codes. The decoding problem for a code C over $\mathbb{F}_q$ of length n means that we want to find one (or all) codeword(s) of C within a prescribed Hamming distance from a given $\mathbf{x} \in \mathbb{F}_q^n$. Another version of the decoding problem asks for a codeword of C that is closest to $\mathbf{x}$ in terms of Hamming distance. A survey of decoding algorithms for AG codes is given in Høholdt and

Pellikaan [56]. More recent contributions to this topic can be found, for instance, in Guruswami and Patthak [47], Guruswami and Sudan [48], and Shokrollahi and Wasserman [113].

Interesting codes can be derived also from algebraic varieties of dimension greater than 1 over finite fields. This idea is mentioned already in the book of Tsfasman and Vlădut [122, Section 3.1.1]. A rather general approach to the construction of such codes is described in Hansen [50]. Constructions using special varieties are discussed, for instance, in the papers of Ghorpade and Tsfasman [42], Lachaud [66], Nogin [102], and Rodier [106].

5.3 Asymptotic Results

As we have noted in Section 5.1, the error-correction capability of a code C is governed by its minimum distance $d(C)$. Just as the information rate $R(C)$ relates the size $|C|$ of the code to its length $n(C)$, it is meaningful to consider the *relative minimum distance* $d(C)/n(C)$. The problem of constructing good codes is then often posed in the following way: for a given error-correction capability (which is suggested by the error rate and hence by physical properties of the communication channel), or equivalently a given relative minimum distance, and a given q, find codes over $\mathbb{F}_q$ for which the information rate is as large as possible.

The asymptotic theory of codes treats this problem for sufficiently long codes. The basic object of study in this theory is the following set of ordered pairs of asymptotic relative minimum distances and asymptotic information rates. For a given prime power q, let U_q be the set of points (δ, R) in the unit square $[0, 1]^2$ for which there exists a sequence $C_1, C_2, \ldots$ of codes over $\mathbb{F}_q$ such that $|C_i| \geq 2$ for all $i \geq 1$, $n(C_i) \to \infty$ as $i \to \infty$, and

$$\lim_{i \to \infty} \frac{d(C_i)}{n(C_i)} = \delta, \qquad \lim_{i \to \infty} R(C_i) = R.$$

In view of the problem posed above, it is reasonable to introduce the following function.

Definition 5.3.1. For a given prime power q, put

$$\alpha_q(\delta) = \sup \{R \in [0, 1] : (\delta, R) \in U_q\} \quad \text{for } 0 \leq \delta \leq 1.$$

Thus, $\alpha_q(\delta)$ can be interpreted as the largest asymptotic information rate that can be achieved for a given asymptotic relative minimum distance δ of codes over $\mathbb{F}_q$ of increasing length.

It can be shown by standard coding-theoretic arguments (see [122, Section 1.3.1]) that α_q is a nonincreasing continuous function on $[0, 1]$ and that

$$U_q = \{(\delta, R) \in [0, 1]^2 : 0 \leq R \leq \alpha_q(\delta)\}.$$

Consequently, the set U_q is completely determined by the function α_q.

The study of the function α_q is a major issue in coding theory. The only known values of α_q are $\alpha_q(0) = 1$ and $\alpha_q(\delta) = 0$ for $(q-1)/q \leq \delta \leq 1$ (see again [122, Section 1.3.1]). The function α_q is not known explicitly on the open interval $(0, (q-1)/q)$. The next best thing is then to give lower bounds on $\alpha_q(\delta)$ for $0 < \delta < (q-1)/q$, so that for a given δ in this range we know at least some asymptotic information rates that can be achieved.

The classical lower bound on α_q can be derived from Proposition 5.1.6. First we introduce the q-ary *entropy function* H_q by $H_q(0) = 0$ and

$$H_q(\delta) = \delta \log_q(q-1) - \delta \log_q \delta - (1-\delta) \log_q(1-\delta) \quad \text{for } 0 < \delta < 1.$$

Note that $H_q(\delta)$ increases from 0 to 1 as δ runs from 0 to $(q-1)/q$.

Theorem 5.3.2 (Asymptotic Gilbert-Varshamov Bound). For any prime power q, we have

$$\alpha_q(\delta) \geq R_{\mathrm{GV}}(q, \delta) := 1 - H_q(\delta) \quad \text{for } 0 \leq \delta \leq \frac{q-1}{q}.$$

Proof. We can assume that $0 < \delta < (q-1)/q$. Put $d_n = \lfloor \delta n \rfloor + 2$ and note that $d_n \leq n$ for sufficiently large n. We have

$$\sum_{i=0}^{d_n - 2} \binom{n-1}{i}(q-1)^i \leq (\lfloor \delta n \rfloor + 1)\binom{n}{\lfloor \delta n \rfloor}(q-1)^{\lfloor \delta n \rfloor}$$

since the largest term of the sum on the left-hand side is the last one. Hence, according to Proposition 5.1.6, a sufficient condition for the existence of a linear $[n, k]$ code over $\mathbb{F}_q$ with $1 \leq k < n$ and minimum distance $\geq d_n$ is

$$(\lfloor \delta n \rfloor + 1)\binom{n}{\lfloor \delta n \rfloor}(q-1)^{\lfloor \delta n \rfloor} < q^{n-k}. \tag{5.4}$$

By Stirling's formula we have

$$\lim_{n \to \infty} \frac{1}{n} \log_q \binom{n}{\lfloor \delta n \rfloor} = -\delta \log_q \delta - (1 - \delta) \log_q (1 - \delta), \qquad (5.5)$$

and so

$$\lim_{n \to \infty} \frac{1}{n} \log_q \left((\lfloor \delta n \rfloor + 1) \binom{n}{\lfloor \delta n \rfloor} (q - 1)^{\lfloor \delta n \rfloor} \right) = H_q(\delta).$$

Now choose a real number $\varepsilon > 0$ with $H_q(\delta) + \varepsilon < 1$ and put

$$k_n = n - \lfloor (H_q(\delta) + \varepsilon) n \rfloor.$$

Then $1 \le k_n < n$ for sufficiently large n and

$$\lim_{n \to \infty} \frac{1}{n} \log_q q^{n - k_n} = \lim_{n \to \infty} \frac{n - k_n}{n} = H_q(\delta) + \varepsilon$$

$$> \lim_{n \to \infty} \frac{1}{n} \log_q \left((\lfloor \delta n \rfloor + 1) \binom{n}{\lfloor \delta n \rfloor} (q - 1)^{\lfloor \delta n \rfloor} \right).$$

Thus, the condition (5.4) is satisfied for sufficiently large n with $k = k_n$, and so for these n there exists a linear $[n, k_n]$ code over $\mathbb{F}_q$ with minimum distance $\ge d_n$. By passing, if necessary, to a subsequence $C_1, C_2, \ldots$ of these codes, we can assume that the limit

$$\delta_0 := \lim_{i \to \infty} \frac{d(C_i)}{n(C_i)}$$

exists. Since $d(C_i) \ge \lfloor \delta n(C_i) \rfloor + 2$, we get $\delta_0 \ge \delta$. Using that α_q is a nonincreasing function, we obtain

$$\alpha_q(\delta) \ge \alpha_q(\delta_0) \ge \lim_{i \to \infty} R(C_i) = \lim_{i \to \infty} \frac{k_{n(C_i)}}{n(C_i)} = 1 - H_q(\delta) - \varepsilon.$$

Letting ε tend to 0, we arrive at the desired result. $\qquad \square$

It is a remarkable fact that one can always achieve the asymptotic Gilbert-Varshamov bound by using sequences of AG codes. This result was established by Xing [134]. For the proof, we first consider a fixed global function field $F/\mathbb{F}_q$ and define the following sets of divisors of F for any integers $m \geq 0$ and $u \geq 0$ and any set $\mathcal{P}$ of rational places of F:

$$\mathcal{A}_m = \{D \in \mathrm{Div}(F) : D \geq 0,\ \deg(D) = m\},$$

$$\mathcal{A}_m(\mathcal{P}) = \left\{ H \in \mathcal{A}_m : H \leq \sum_{P \in \mathcal{P}} P \right\},$$

$$\mathcal{M}_{u,m}(\mathcal{P}) = \{D + H : D \in \mathcal{A}_u,\ H \in \mathcal{A}_m(\mathcal{P})\}.$$

Then we establish the following refinement of Theorems 5.2.2 and 5.2.7. We recall that $h(F)$ denotes the divisor class number of F.

Proposition 5.3.3. Let $F/\mathbb{F}_q$ be a global function field of genus g with $N(F) \geq 1$. Let m, n, and u be integers with $1 \leq m \leq n \leq N(F)$, $u \geq 0$, and $m + u \geq g$. Let $\mathcal{P}$ be a set of rational places of F with $|\mathcal{P}| = n$ such that $|\mathcal{M}_{u,m}(\mathcal{P})| < h(F)$. Then there exists an AG code, which is a linear $[n, k, d]$ code over $\mathbb{F}_q$ with

$$k \geq m + u + 1 - g, \qquad d \geq n - m + 1.$$

Proof. Recall from Section 4.1 that there are exactly $h(F)$ divisor classes of F of degree $m + u$. Since $|\mathcal{M}_{u,m}(\mathcal{P})| < h(F)$, there exists a divisor G of F with $\deg(G) = m + u$ that is not equivalent to any divisor in $\mathcal{M}_{u,m}(\mathcal{P})$. Moreover, by an argument in the proof of Theorem 5.2.7 we can assume that $\mathrm{supp}(G) \cap \mathcal{P} = \varnothing$.

We claim that $\mathcal{L}(G - \sum_{P \in \mathcal{J}} P) = \{0\}$ for any $\mathcal{J} \subseteq \mathcal{P}$ with $|\mathcal{J}| = m$. Suppose, on the contrary, that for some $\mathcal{J} \subseteq \mathcal{P}$ with $|\mathcal{J}| = m$ we have $\mathcal{L}(G - \sum_{P \in \mathcal{J}} P) \neq \{0\}$. Choose a nonzero $x \in \mathcal{L}(G - \sum_{P \in \mathcal{J}} P)$. Then

$$D := G - \sum_{P \in \mathcal{J}} P + \mathrm{div}(x) \geq 0.$$

But G is equivalent to $D + \sum_{P \in \mathcal{J}} P \in \mathcal{M}_{u,m}(\mathcal{P})$, and this is a contradiction to the choice of G.

With $\mathcal{P} = \{P_1, \ldots, P_n\}$ we consider now the AG code $C(P_1, \ldots, P_n; G)$. For a nonzero $f \in \mathcal{L}(G)$, let w be the weight of $\psi(f)$ in (5.3). Then $f \in \mathcal{L}(G - \sum_{P \in \mathcal{J}} P)$ for some $\mathcal{J} \subseteq \mathcal{P}$ with $|\mathcal{J}| = n - w$. By what we have proved above, we must have $n - w < m$, and so $w \geq n - m + 1$. This shows not only the desired lower bound on d, but also that ψ is injective. Hence, $k = \ell(G)$, and the rest follows from Riemann's theorem. □

For an integer $b \geq 0$ and a set $\mathcal{P}$ of rational places of F, we define

$$\mathcal{A}_b^{(\mathcal{P})} = \{D \in \mathcal{A}_b : \operatorname{supp}(D) \cap \mathcal{P} = \varnothing\}.$$

For fixed F, it is clear that the cardinality of $\mathcal{A}_b^{(\mathcal{P})}$ depends only on b and the cardinality of $\mathcal{P}$, and not on the actual elements of $\mathcal{P}$. Hence, we can write $|\mathcal{A}_b^{(\mathcal{P})}| =: A_b^{(s)}$ if $|\mathcal{P}| = s$.

Lemma 5.3.4. Let $F/\mathbb{F}_q$ be a global function field of genus $g \geq 1$. Then for any integers $b \geq 0$ and $0 \leq s \leq N(F)$ we have

$$A_b^{(s)} \leq \frac{h(F)}{q^{g-b}} \left(\frac{2gq^{1/2}}{q^{1/2} - 1} + \frac{(s-1)q}{q-1} \right) \left(1 - \frac{1}{q} \right)^{s-1}.$$

Proof. As in Definition 4.1.1, we denote the cardinality of $\mathcal{A}_h$ by $A_b(F)$. By writing any positive divisor D of F in the form

$$D = \sum_{P \in \mathbf{P}_F} v_P(D)P = \sum_{P \in \mathcal{P}} v_P(D)P + \sum_{P \in \mathbf{P}_F \setminus \mathcal{P}} v_P(D)P$$

with a chosen set $\mathcal{P}$ of rational places of F with $|\mathcal{P}| = s$, we see that

$$A_b(F) = \sum_{j=0}^{b} A_{b-j}^{(s)} \left| \left\{ E = \sum_{P \in \mathcal{P}} v_P(E)P : E \geq 0, \deg(E) = j \right\} \right|.$$

For a formal power series $B(t)$ over $\mathbb{R}$ in the variable t, we write $[B(t)]_j$ for the coefficient of t^j in $B(t)$. Then it follows that

$$A_b(F) = \sum_{j=0}^{b} A_{b-j}^{(s)} \left[\left(\sum_{k=0}^{\infty} t^k \right)^s \right]_j.$$

In view of Theorem 4.1.6(ii), we can then write the zeta function $Z(F, t)$ of F as

$$Z(F, t) = \sum_{b=0}^{\infty} A_b(F) t^b = \sum_{b=0}^{\infty} \left(\sum_{j=0}^{b} A_{b-j}^{(s)} [(1-t)^{-s}]_j \right) t^b$$

$$= \left(\sum_{b=0}^{\infty} A_b^{(s)} t^b \right) (1-t)^{-s} =: Z^{(s)}(F, t)(1-t)^{-s},$$

and so

$$Z^{(s)}(F, t) = Z(F, t)(1-t)^s \quad \text{for } 0 \le s \le N(F). \tag{5.6}$$

By (4.2) and the proof of Lemma 4.1.10, we have

$$A_b(F) = \frac{h(F)}{q-1}(q^{b+1-g} - 1) \quad \text{for } b \ge 2g - 1. \tag{5.7}$$

Thus, for $0 \le s \le N(F)$ and $b \ge 2g + s - 1$, we get from (5.6),

$$A_b^{(s)} = [Z^{(s)}(F, t)]_b = \sum_{j=0}^{s} [(1-t)^s]_j [Z(F, t)]_{b-j}$$

$$= \sum_{j=0}^{s} (-1)^j \binom{s}{j} A_{b-j}(F) = \frac{h(F)}{q-1} \sum_{j=0}^{s} (-1)^j \binom{s}{j}(q^{b-j+1-g} - 1)$$

$$= \frac{h(F)}{q-1} q^{b+1-g} \sum_{j=0}^{s} (-1)^j \binom{s}{j} q^{-j} = h(F) q^{b-g} \left(1 - \frac{1}{q}\right)^{s-1}.$$

This proves the result of the lemma for these values of s and b, so that it remains to consider $0 \le s \le N(F)$ and $0 \le b \le 2g + s - 2$. By the identity we have just shown, we can write

$$
\sum_{b=0}^{2g+s-2} A_b^{(s)} t^b = Z(F,t)(1-t)^s - \sum_{b=2g+s-1}^{\infty} A_b^{(s)} t^b
$$

$$
= Z(F,t)(1-t)^s - h(F)q^{-g}\left(1-\frac{1}{q}\right)^{s-1} \sum_{b=2g+s-1}^{\infty} (qt)^b
$$

$$
= Z(F,t)(1-t)^s - \frac{h(F)(q-1)^{s-1}q^g t^{2g+s-1}}{1-qt}.
$$

Now we use Theorem 4.1.11 to obtain

$$
\sum_{b=0}^{2g+s-2} A_b^{(s)} t^b = \frac{L(F,t)(1-t)^{s-1} - h(F)(q-1)^{s-1}q^g t^{2g+s-1}}{1-qt}.
$$

We can evaluate the last expression at $t = \frac{1}{q}$ by differentiating numerator and denominator at $t = \frac{1}{q}$. Note that

$$
L\left(F,\frac{1}{q}\right) = \frac{L(F,1)}{q^g} = \frac{h(F)}{q^g}
$$

by Theorems 4.1.13 and 4.1.11. Thus, we get

$$
\sum_{b=0}^{2g+s-2} \frac{A_b^{(s)}}{q^b} = \frac{1}{-q}\left(L'\left(F,\frac{1}{q}\right)\left(1-\frac{1}{q}\right)^{s-1} - \frac{h(F)(s-1)}{q^g}\left(1-\frac{1}{q}\right)^{s-2}\right.
$$

$$
\left. - \frac{h(F)(q-1)^{s-1}(2g+s-1)}{q^{g+s-2}}\right).
$$

The computation of $L'(F,\frac{1}{q})$ proceeds by logarithmic differentiation of (4.6). This yields

$$
L'\left(F,\frac{1}{q}\right) = \frac{h(F)}{q^{g-1}} \sum_{i=1}^{2g} \frac{\omega_i}{\omega_i - q}.
$$

Now $|\omega_i| = q^{1/2}$ for $1 \le i \le 2g$ by Theorem 4.2.3; hence,

$$|L'\left(F, \frac{1}{q}\right)| \le \frac{2gh\,(F)}{\left(q^{1/2} - 1\right) q^{g-1}}.$$

This implies that

$$\sum_{b=0}^{2g+s-2} \frac{A_b^{(s)}}{q^b} \le \frac{h(F)}{q^g}\left(\frac{2g}{q^{1/2} - 1} + \frac{s - 1}{q - 1} + 2g + s - 1\right)\left(1 - \frac{1}{q}\right)^{s-1}$$

$$= \frac{h(F)}{q^g}\left(\frac{2gq^{1/2}}{q^{1/2} - 1} + \frac{(s - 1)q}{q - 1}\right)\left(1 - \frac{1}{q}\right)^{s-1}.$$

For each integer b with $0 \le b \le 2g + s - 2$, this yields

$$\frac{A_b^{(s)}}{q^b} \le \frac{h(F)}{q^g}\left(\frac{2gq^{1/2}}{q^{1/2} - 1} + \frac{(s - 1)q}{q - 1}\right)\left(1 - \frac{1}{q}\right)^{s-1},$$

which is the desired bound.   □

We can now bound the cardinality of the sets $\mathcal{M}_{u,m}(\mathcal{P})$ defined earlier, in the case where $|\mathcal{P}| = N(F)$.

Lemma 5.3.5. Let $F/\mathbb{F}_q$ be a global function field of genus $g \ge 1$ and put $n = N(F)$. Let $\mathcal{P}$ be the set of all rational places of F. Then for any integers $u \ge 0$ and $0 \le m \le n$ we have

$$|\mathcal{M}_{u,m}(\mathcal{P})| \le \frac{(7g + 2n)h(F)}{q^{g+n-m-u-1}} \sum_{j=0}^{\min(u,n-m)} \binom{n}{m + j}(q - 1)^{n-m-j-1}.$$

Proof. The set $\mathcal{M}_{u,m}(\mathcal{P})$ is the disjoint union of the sets

$$\mathcal{M}_{u,m,j}(\mathcal{P}) = \{D + H : D \in \mathcal{A}_{u-j}, H \in \mathcal{A}_{m+j}(\mathcal{P}),$$

$$\text{supp}(D) \cap (\mathcal{P} \setminus \text{supp}(H)) = \varnothing\}$$

with $j = 0, 1, \ldots, u$, and these sets are empty for $j > n - m$.

Therefore,

$$|\mathcal{M}_{u,m}(\mathcal{P})| = \sum_{j=0}^{\min(u,n-m)} |\mathcal{M}_{u,m,j}(\mathcal{P})| = \sum_{j=0}^{\min(u,n-m)} \binom{n}{m+j} A_{u-j}^{(n-m-j)}$$

since $|\mathcal{A}_{m+j}(\mathcal{P})| = \binom{n}{m+j}$ for $0 \le j \le n - m$. By using the simplified bound

$$A_b^{(s)} \le \frac{h(F)}{q^{g-b}}(7g+2n)\left(1 - \frac{1}{q}\right)^{s-1}$$

deduced from Lemma 5.3.4, we infer the desired result.   □

Corollary 5.3.6. Let $F/\mathbb{F}_q$ be a global function field of genus $g \ge 1$ with $N(F) \ge 1$. Let m and u be integers with $1 \le m \le N(F)$ and $m + u \ge g$. Suppose that

$$\left(7g + 2N(F)\right) \sum_{j=0}^{\min(u,N(F)-m)} \binom{N(F)}{m+j}(q-1)^{N(F)-m-j-1} < q^{g+N(F)-m-u-1}.$$

$$(5.8)$$

Then there exists an AG code, which is a linear $[n, k, d]$ code over $\mathbb{F}_q$ with

$$n = N(F), \qquad k \ge m + u + 1 - g, \qquad d \ge n - m + 1.$$

Proof. For $u < 0$ the result follows by applying Theorem 5.2.7 with $n = N(F)$ and $\deg(G) = m + u$. For $u \ge 0$ it follows from Proposition 5.3.3 and Lemma 5.3.5.   □

After these preparations, we are now ready to prove with Xing [134] that the asymptotic Gilbert-Varshamov bound $R_{GV}(q, \delta)$ in Theorem 5.3.2 can be obtained by using sequences of AG codes.

Theorem 5.3.7. For any prime power q and any real number δ with $0 \le \delta \le (q-1)/q$, there exists a sequence $C_1, C_2, \ldots,$ of AG codes over $\mathbb{F}_q$ with $n(C_i) \to \infty$ as $i \to \infty$ which yields

$$\alpha_q(\delta) \ge R_{GV}(q, \delta).$$

Proof. The result is trivial for $\delta = (q-1)/q$ since $R_{GV}(q, (q-1)/q) = 0$. Thus, we can assume that $0 \leq \delta < (q-1)/q$.

Since $A(q) > 0$ by Remark 4.3.8, there exists a sequence $F_1/\mathbb{F}_q$, $F_2/\mathbb{F}_q, \ldots$ of global function fields with $g_i := g(F_i) \to \infty$ as $i \to \infty$ and such that $n_i := N(F_i)$ satisfies

$$\lim_{i \to \infty} \frac{n_i}{g_i} > 0. \tag{5.9}$$

Choose a real number $\varepsilon > 0$ with $H_q(\delta) + \varepsilon < 1$. For $i = 1, 2, \ldots$ put

$$a_i = \lfloor \delta n_i \rfloor, \quad b_i = n_i - \lfloor (H_q(\delta) + \varepsilon) n_i \rfloor.$$

Then for sufficiently large i we have $0 \leq a_i < (q-1)n_i/q$ and $1 \leq b_i < n_i$. We also define

$$m_i = n_i - a_i, \quad u_i = a_i + b_i + g_i - n_i - 1.$$

Then for sufficiently large i we have $1 \leq m_i \leq n_i$ and $m_i + u_i = b_i + g_i - 1 \geq g_i$.

We claim that the condition (5.8) holds with $F = F_i$, $m = m_i$, and $u = u_i$ for sufficiently large i. With these choices, let E_i be the expression on the left-hand side of (5.8). Note that

$$E_i \leq (7g_i + 2n_i) \sum_{j=0}^{a_i} \binom{n_i}{n_i - a_i + j} (q-1)^{a_i - j - 1}.$$

Using $a_i < (q-1)n_i/q$, it is easily checked that

$$\binom{n_i}{n_i - a_i + j} (q-1)^{a_i - j - 1}$$

is a nonincreasing function of $j = 0, 1, \ldots, a_i$. Therefore,

$$E_i \leq (7g_i + 2n_i)(a_i + 1) \binom{n_i}{a_i} (q-1)^{a_i - 1}.$$

Recalling that $a_i = \lfloor \delta n_i \rfloor$ and using (5.5) and (5.9), we obtain

$$\limsup_{i \to \infty} \frac{1}{n_i} \log_q E_i \leq H_q(\delta).$$

Furthermore, with the choices we have made, the right-hand side of (5.8) becomes $q^{n_i - b_i}$, and we have

$$\lim_{i \to \infty} \frac{1}{n_i} \log_q q^{n_i - b_i} = H_q(\delta) + \varepsilon.$$

Thus, (5.8) is indeed satisfied for sufficiently large i.

Consequently, for sufficiently large i, Corollary 5.3.6 yields an AG code, which is a linear $[n_i, k_i, d_i]$ code over $\mathbb{F}_q$ with $k_i \geq b_i$ and $d_i \geq a_i + 1$. By passing, if necessary, to a subsequence $C_1, C_2, \ldots$ of these AG codes, we can assume that the limits

$$\delta_0 := \lim_{i \to \infty} \frac{d(C_i)}{n(C_i)} \quad \text{and} \quad R_0 := \lim_{i \to \infty} R(C_i)$$

exist. From the lower bounds on d_i and k_i we infer that $\delta_0 \geq \delta$ and $R_0 \geq 1 - H_q(\delta) - \varepsilon$. Using that α_q is a nonincreasing function, we obtain

$$\alpha_q(\delta) \geq \alpha_q(\delta_0) \geq R_0 \geq 1 - H_q(\delta) - \varepsilon = R_{\mathrm{GV}}(q, \delta) - \varepsilon.$$

We complete the proof by letting ε tend to 0. $\qquad\square$

It was a milestone in coding theory when it was discovered that, for certain q and δ, sequences of AG codes can even be used to beat the asymptotic Gilbert-Varshamov bound $R_{\mathrm{GV}}(q, \delta)$. The basic tool for this purpose is the following result due to Tsfasman, Vlădut, and Zink [123]. Here $A(q)$ is the number that was introduced in Definition 4.3.6.

Theorem 5.3.8 (TVZ Bound). For any prime power q, we have

$$\alpha_q(\delta) \geq R_{\mathrm{TVZ}}(q, \delta) := 1 - \frac{1}{A(q)} - \delta \quad \text{for } 0 \leq \delta \leq 1.$$

Proof. We can assume that $A(q) > 1$ and $0 \leq \delta \leq 1 - \frac{1}{A(q)}$, for otherwise the result is trivial. Let $F_1/\mathbb{F}_q, F_2/\mathbb{F}_q, \ldots$ be a sequence of global function fields such that $g_i := g(F_i)$ and $n_i := N(F_i)$ satisfy

$$\lim_{i \to \infty} g_i = \infty \quad \text{and} \quad \lim_{i \to \infty} \frac{n_i}{g_i} = A(q).$$

For sufficiently large i we can choose integers r_i with $g_i \le r_i < n_i$ and

$$\lim_{i \to \infty} \frac{r_i}{n_i} = 1 - \delta.$$

For such i, Theorem 5.2.7 yields an AG code, which is a linear $[n_i, k_i, d_i]$ code over $\mathbb{F}_q$ with

$$k_i \ge r_i + 1 - g_i \quad \text{and} \quad d_i \ge n_i - r_i.$$

By passing, if necessary, to a subsequence, we can assume that the limits

$$\delta_0 := \lim_{i \to \infty} \frac{d_i}{n_i} \quad \text{and} \quad R_0 := \lim_{i \to \infty} \frac{k_i}{n_i}$$

exist. We have

$$\delta_0 \ge \delta \quad \text{and} \quad R_0 \ge 1 - \frac{1}{A(q)} - \delta.$$

It follows that

$$\alpha_q(\delta) \ge \alpha_q(\delta_0) \ge R_0 \ge 1 - \frac{1}{A(q)} - \delta$$

since the function α_q is nonincreasing. $\square$

If q is a square, then we know from Remark 4.3.8 that $A(q) = q^{1/2} - 1$. It is now simply a matter of comparing the two explicitly given functions $R_{\mathrm{GV}}(q, \delta)$ and $R_{\mathrm{TVZ}}(q, \delta)$ on the interval $[0, (q - 1)/q]$ to determine when sequences of AG codes can yield an improvement on the asymptotic Gilbert-Varshamov bound.

Theorem 5.3.9. Let $q \ge 49$ be the square of a prime power. Then there exists an open interval $(\delta_1, \delta_2) \subseteq [0, (q - 1)/q]$ containing $(q - 1)/(2q - 1)$ such that

$$R_{\mathrm{TVZ}}(q, \delta) > R_{\mathrm{GV}}(q, \delta) \quad \text{for all } \delta \in (\delta_1, \delta_2).$$

Proof. It suffices to prove the inequality for $\delta_0 = (q-1)/(2q-1)$. Note that

$$H_q(\delta_0) - \delta_0 = \log_q \left(2 - \frac{1}{q} \right).$$

Thus, it remains to show that

$$\log_q \left(2 - \frac{1}{q} \right) > \frac{1}{q^{1/2} - 1}.$$

It is easily checked that this holds for all $q \geq 49$.   □

Remark 5.3.10. According to a result of Bezerra, Garcia, and Stichtenoth [7], we have

$$A(q) \geq \frac{2(q^{2/3} - 1)}{q^{1/3} + 2}$$

whenever q is the cube of a prime power. This was shown earlier by Zink [138] in the special case where q is the cube of a prime. To establish a result like Theorem 5.3.9 for cubes q, it suffices to prove that

$$\log_q \left(2 - \frac{1}{q} \right) > \frac{q^{1/3} + 2}{2(q^{2/3} - 1)}.$$

It is an easy exercise to verify that this holds for all $q \geq 343$. Thus, the analog of Theorem 5.3.9 holds whenever $q \geq 343$ is the cube of a prime power. More generally, Niederreiter and Xing [99] (see also [100, Theorem 6.2.8]) have shown that a result like Theorem 5.3.9 holds for all sufficiently large composite nonsquare prime powers q. It is not known whether such a result holds also for all sufficiently large primes q.

The lower bounds on $\alpha_q(\delta)$ that we have established so far, namely Theorems 5.3.2, 5.3.7, and 5.3.8, have been proved by using sequences of *linear* codes. It is an important fact that the TVZ bound in Theorem 5.3.8 can be improved by using codes that are not necessarily linear. Indeed,

Niederreiter and Özbudak [94] showed that

$$\alpha_q(\delta) \geq R_{\text{NÖ}}(q, \delta) := 1 - \frac{1}{A(q)} - \delta + \log_q \left(1 + \frac{1}{q^3}\right) \quad \text{for } 0 \leq \delta \leq 1.$$
$$(5.10)$$

This result followed a number of earlier improvements on the TVZ bound obtained by Vlăduţ [124] (see also [122, Chapter 3.4]), Elkies [31], and Xing [131, 133]. The original proof of (5.10) is quite involved. We present a considerably simpler proof of this inequality, which was given by Stichtenoth and Xing [118], albeit for a slightly restricted range of the parameter δ (see Theorem 5.3.13 below).

We need a few preparations for this proof. Let $F/\mathbb{F}_q$ be a global function field of genus g with $N(F) \geq 1$ and let r be an integer with $1 \leq r \leq N(F)$. Let $P_1, \ldots, P_r$ be r distinct rational places of F and let $D \geq 0$ be a divisor of F with $\text{supp}(D) \cap \{P_1, \ldots, P_r\} = \varnothing$. Let $G = \sum_{i=1}^r m_i P_i$ with positive integers $m_1, \ldots, m_r$ and put

$$W_D(G) := \{f \in \mathcal{L}(D + G) : v_{P_i}(f) = -m_i \quad \text{for } 1 \leq i \leq r\}.$$

Lemma 5.3.11. If in addition to the above conditions we have $m := \deg(D) \geq 2g - 1$, then

$$|W_D(G)| = q^{m+s-g+1} \left(1 - \frac{1}{q}\right)^r,$$

where $s = \deg(G)$.

Proof. It is obvious that

$$W_D(G) = \mathcal{L}(D + G) \setminus \left(\bigcup_{i=1}^r \mathcal{L}(D + G - P_i)\right).$$

By the Riemann-Roch theorem we have

$$|\mathcal{L}(D + G)| = q^{m+s-g+1}.$$

Furthermore, the inclusion-exclusion principle in combinatorics yields

$$
\left| \bigcup_{i=1}^{r} \mathcal{L}(D + G - P_i) \right| = \sum_{k=1}^{r}(-1)^{k+1} \sum_{1 \leq i_1 < i_2 < \cdots < i_k \leq r} \left| \bigcap_{j=1}^{k} \mathcal{L}(D + G - P_{i_j}) \right|
$$

$$
= \sum_{k=1}^{r}(-1)^{k+1} \sum_{1 \leq i_1 < i_2 < \cdots < i_k \leq r} \left| \mathcal{L}\left(D + G - \sum_{j=1}^{k} P_{i_j} \right) \right|
$$

$$
= \sum_{k=1}^{r}(-1)^{k+1} \binom{r}{k} q^{m+s-g-k+1}
$$

$$
= q^{m+s-g+1} \left(1 - \left(1 - \frac{1}{q} \right)^{r} \right).
$$

The result of the lemma follows now immediately. $\qquad \square$

The basic construction of codes in [118] is a variant of the construction of AG codes and proceeds as follows. Let $F/\mathbb{F}_q$ be a global function field with $N(F) \geq 2$. Let n be an integer with $1 \leq n \leq N(F) - 1$ and choose $n + 1$ distinct rational places $P_0, P_1, \ldots, P_n$ of F. Put $\mathcal{P} = \{P_1, \ldots, P_n\}$. Furthermore, let $m, r,$ and s be integers with $m \geq 1$ and $1 \leq r \leq \min(n, s)$. Recall the notation $\mathcal{A}_s$ for the set of positive divisors of F of degree s. We put

$$
\mathcal{G}(\mathcal{P}; r, s) = \{G \in \mathcal{A}_s : \operatorname{supp}(G) \subseteq \mathcal{P}, \ |\operatorname{supp}(G)| = r\}.
$$

Now we define

$$
S(m P_0; \mathcal{P}; r, s) = \bigcup_{G \in \mathcal{G}(\mathcal{P};r,s)} W_{m P_0}(G).
$$

Note that $S(m P_0; \mathcal{P}; r, s)$ is the disjoint union of the sets $W_{m P_0}(G)$ with $G \in \mathcal{G}(\mathcal{P}; r, s)$. Thus, we can define a map

$$
\phi : S(m P_0; \mathcal{P}; r, s) \rightarrow \mathbb{F}_q^n
$$

in the following way. Take $f \in S(m P_0; \mathcal{P}; r, s)$, then $f \in W_{m P_0}(G)$ for a uniquely determined $G \in \mathcal{G}(\mathcal{P}; r, s)$. We set

$$
\phi(f) = (c_1(f), \ldots, c_n(f)) \in \mathbb{F}_q^n,
$$

where for $i = 1, \ldots, n$ we put

$$
c_i(f) = \begin{cases} f(P_i) & \text{if } P_i \notin \operatorname{supp}(G), \\ 0 & \text{if } P_i \in \operatorname{supp}(G). \end{cases}
$$

The desired code is now $C(m P_0; \mathcal{P}; r, s) := \phi(S(m P_0; \mathcal{P}; r, s)) \subseteq \mathbb{F}_q^n$. The following result yields information on the parameters of this code.

Proposition 5.3.12. Let $F/\mathbb{F}_q$ be a global function field of genus g. Let m, n, r, and s be integers with

$$
m \geq \max(2g - 1, 1), \qquad 1 \leq r \leq s, \qquad n - m - 2r - 2s \geq 1.
$$

Assume also that $N(F) \geq n + 1$. Then with the notation above, $C := C(m P_0; \mathcal{P}; r, s)$ is a code over $\mathbb{F}_q$ of length n with

$$
|C| = q^{m+s-g+1} \left(1 - \frac{1}{q}\right)^r \binom{n}{r}\binom{s-1}{r-1},
$$

$$
d(C) \geq n - m - 2r - 2s.
$$

Proof. Take $f_1, f_2 \in S(m P_0; \mathcal{P}; r, s)$ with $f_1 \neq f_2$. Then $f_j \in W_{m P_0}(G_j)$ for $j = 1, 2$ and some $G_1, G_2 \in \mathcal{G}(\mathcal{P}; r, s)$. It follows that $f_1 - f_2 \in \mathcal{L}(m P_0 + G_1 + G_2)$. Consider the set

$$
\mathcal{Z} = \{P \in \mathcal{P} : P \notin \operatorname{supp}(G_1) \cup \operatorname{supp}(G_2) \text{ and } f_1(P) = f_2(P)\}.
$$

Then $f_1 - f_2 \in \mathcal{L}(m P_0 + G_1 + G_2 - \sum_{P \in \mathcal{Z}} P)$, and since $f_1 - f_2 \neq 0$, this implies

$$
\deg\left(m P_0 + G_1 + G_2 - \sum_{P \in \mathcal{Z}} P\right) = m + 2s - |\mathcal{Z}| \geq 0
$$

by Corollary 3.4.4. Furthermore, the weight w of $\phi(f_1) - \phi(f_2)$ satisfies

$$
w = n - |\{1 \leq i \leq n : c_i(f_1) = c_i(f_2)\}|,
$$

and we have

$$|\{1 \leq i \leq n : c_i(f_1) = c_i(f_2)\}| \leq 2r + |\mathcal{Z}|.$$

Therefore, $w \geq n - 2r - |\mathcal{Z}| \geq n - m - 2r - 2s \geq 1$. This shows not only the desired lower bound on $d(C)$, but also that ϕ is injective. Hence, by Lemma 5.3.11,

$$|C| = |S(m\,P_0; \mathcal{P}; r, s)| = q^{m+s-g+1} \left(1 - \frac{1}{q}\right)^r |\mathcal{G}(\mathcal{P}; r, s)|$$

$$= q^{m+s-g+1} \left(1 - \frac{1}{q}\right)^r \binom{n}{r} \binom{s-1}{r-1}$$

and the proof is complete. $\qquad\qquad\qquad\qquad\qquad\qquad\qquad$ □

Theorem 5.3.13. For any prime power q and any real number δ with

$$0 \leq \delta \leq 1 - \frac{2}{A(q)} - \frac{4q - 2}{(q-1)(q^3+1)},$$

we have

$$\alpha_q(\delta) \geq R_{\text{NÖ}}(q, \delta) = 1 - \frac{1}{A(q)} - \delta + \log_q\left(1 + \frac{1}{q^3}\right).$$

Proof. Since the function α_q is continuous, we can assume that

$$0 < \delta < 1 - \frac{2}{A(q)} - \frac{4q - 2}{(q-1)(q^3+1)}.$$

Choose a sequence $F_1/\mathbb{F}_q, F_2/\mathbb{F}_q, \ldots$ of global function fields such that $g_i := g(F_i) \to \infty$ as $i \to \infty$ and

$$\lim_{i \to \infty} \frac{N(F_i)}{g_i} = A(q).$$

Then $g_i > 0$ and $N(F_i) \geq 2$ for sufficiently large i. For $i = 1, 2, \ldots$ put

$$n_i = N(F_i) - 1, \qquad s_i = \left\lfloor \frac{qn_i}{(q-1)(q^3+1)} \right\rfloor, \qquad r_i = \left\lfloor \frac{(q-1)s_i}{q} \right\rfloor.$$

Then the condition $1 \leq r_i \leq s_i$ is satisfied for sufficiently large i. Furthermore, we put

$$m_i = n_i - \lfloor \delta n_i \rfloor - 2r_i - 2s_i.$$

Then

$$\lim_{i \to \infty} \frac{m_i}{n_i} = 1 - \delta - \frac{2}{q^3 + 1} - \frac{2q}{(q-1)(q^3+1)}$$

$$= 1 - \delta - \frac{4q - 2}{(q-1)(q^3+1)} > \frac{2}{A(q)} = \lim_{i \to \infty} \frac{2g_i}{n_i},$$

and so $m_i \geq 2g_i > 0$ for sufficiently large i. Clearly, we also have $n_i - m_i - 2r_i - 2s_i = \lfloor \delta n_i \rfloor \geq 1$ for sufficiently large i. Thus, for these i can we apply Proposition 5.3.12. As usual, we can assume without loss of generality that for the resulting sequence $C_{i_0}, C_{i_0+1}, \ldots$ of codes over $\mathbb{F}_q$ (with a suitable index i_0) the limit

$$\delta_0 := \lim_{i \to \infty} \frac{d(C_i)}{n(C_i)}$$

exists. Note that for $i \geq i_0$, Proposition 5.3.12 yields $n(C_i) = n_i$ as well as

$$|C_i| = q^{m_i+s_i-g_i+1} \left(1 - \frac{1}{q}\right)^{r_i} \binom{n_i}{r_i} \binom{s_i - 1}{r_i - 1},$$

$$d(C_i) \geq n_i - m_i - 2r_i - 2s_i.$$

It follows that

$$\delta_0 \geq \lim_{i \to \infty} \frac{n_i - m_i - 2r_i - 2s_i}{n_i} = \lim_{i \to \infty} \frac{\lfloor \delta n_i \rfloor}{n_i} = \delta.$$

Next we compute the asymptotic information rate. We have

$$\lim_{i \to \infty} R(C_i) = \lim_{i \to \infty} \frac{\log_q |C_i|}{n_i}$$

$$= \lim_{i \to \infty} \frac{m_i + s_i - g_i + 1 + r_i \log_q(1 - q^{-1})}{n_i}$$

$$+ \lim_{i \to \infty} \frac{1}{n_i} \log_q \binom{n_i}{r_i} + \lim_{i \to \infty} \frac{1}{n_i} \log_q \binom{s_i}{r_i}.$$

The last three limits are easily obtained. First of all, we have

$$\lim_{i \to \infty} \frac{m_i + s_i - g_i + 1 + r_i \log_q(1 - q^{-1})}{n_i}$$

$$= 1 - \frac{1}{A(q)} - \delta - \frac{q}{(q-1)(q^3+1)} - \frac{1}{q^3+1} \log_q \frac{q^3}{q-1}.$$

Next we use (5.5) to obtain

$$\lim_{i \to \infty} \frac{1}{n_i} \log_q \binom{n_i}{r_i} = \frac{1}{q^3+1} \log_q(q^3+1) + \frac{q^3}{q^3+1} \log_q \left(1 + \frac{1}{q^3}\right).$$

Finally, again by (5.5),

$$\lim_{i \to \infty} \frac{1}{n_i} \log_q \binom{s_i}{r_i} = \lim_{i \to \infty} \frac{s_i}{n_i} \cdot \frac{1}{s_i} \log_q \binom{s_i}{r_i}$$

$$= \frac{q}{(q-1)(q^3+1)} \left(1 - \frac{q-1}{q} \log_q(q-1)\right)$$

$$= \frac{q}{(q-1)(q^3+1)} - \frac{1}{q^3+1} \log_q(q-1).$$

Adding up these limits, we get

$$\lim_{i \to \infty} R(C_i) = 1 - \frac{1}{A(q)} - \delta + \log_q \left(1 + \frac{1}{q^3}\right).$$

It follows that

$$\alpha_q(\delta) \geq \alpha_q(\delta_0) \geq 1 - \frac{1}{A(q)} - \delta + \log_q \left(1 + \frac{1}{q^3}\right)$$

since the function α_q is nonincreasing.   □

Remark 5.3.14. The bound (5.10) is by no means the last word on improvements of the TVZ bound. More recently, Niederreiter and Özbudak [95], [96] obtained improvements on (5.10), though these improvements are not of a global character, that is, they are not valid on the whole interval $(0, (q - 1)/q)$ that is relevant for the asymptotic theory of codes, but only on certain subintervals thereof. The proofs in [95] and [96] are too complicated to be reproduced here. Results like (5.10) or the improved bounds in [95] and [96] are of interest since they yield larger subintervals of $(0, (q - 1)/q)$ than the TVZ bound on which we can beat the asymptotic Gilbert-Varshamov bound. The approach by Stichtenoth and Xing [118] was refined by Maharaj [76], but his results are superseded by [96].

5.4 NXL and XNL Codes

Results like those in Theorem 5.3.9 and Remark 5.3.10 demonstrate that AG codes yields excellent linear codes over $\mathbb{F}_q$ for certain large values of q. However, AG codes are practically useless for small values of q. An easy way to see this is to consider the TVZ bound in Theorem 5.3.8 and to note that this result produces nontrivial information only when $A(q) > 1$. On the other hand, we deduce from Theorem 4.3.7 that $A(q) \leq 1$ whenever $q \leq 4$. Thus, one cannot expect good long AG codes over $\mathbb{F}_q$ in the important cases $q = 2, 3$, and 4.

The intrinsic reason for the above problem is that the construction of AG codes requires rational places of a global function field over $\mathbb{F}_q$ and that there are just too few rational places relative to the genus for small q. A possible remedy is to devise constructions of linear codes that use also places of higher degree. In this section we discuss two such constructions.

The first construction is due to Niederreiter, Xing, and Lam [101] and leads to the family of NXL codes. We present this construction in a somewhat more general form than in [101]. Let $F/\mathbb{F}_q$ be a global function field of genus g. Let $G_1, \ldots, G_r$ be positive divisors of F with pairwise disjoint supports and with

$s_i := \deg(G_i) \geq 1$ for $1 \leq i \leq r$. Put $n = \sum_{i=1}^{r} s_i$, which will be the length of the code to be constructed, and assume that $n > g$. Let E be a positive divisor of F with $\operatorname{supp}(E) \cap \operatorname{supp}(G_i) = \varnothing$ for $1 \leq i \leq r$. Furthermore, let D be a divisor of F with $\ell(D) = \deg(D) + 1 - g$, that is, for which we have equality in Riemann's theorem. For instance, this holds if $\deg(D) \geq 2g - 1$, according to Theorem 3.6.14. Assume also that $1 \leq \deg(E - D) \leq n - g$.

We observe that $\ell(D + G_i) = \ell(D) + s_i$ for $1 \leq i \leq r$ by Corollary 3.5.2. For each $i = 1, \ldots, r$, we choose an $\mathbb{F}_q$-basis

$$\{f_{i,j} + \mathcal{L}(D) \ : \ 1 \leq j \leq s_i\}$$

of the factor space $\mathcal{L}(D + G_i)/\mathcal{L}(D)$. The n-dimensional factor space $\mathcal{L}(D + \sum_{i=1}^{r} G_i)/\mathcal{L}(D)$ has then the $\mathbb{F}_q$-basis

$$\{f_{i,j} + \mathcal{L}(D) \ : \ 1 \leq j \leq s_i, \ 1 \leq i \leq r\},$$

which we order in a lexicographic manner. The latter basis property is proved by noting that if $\sum_{i=1}^{r} h_i \in \mathcal{L}(D)$ with $h_i \in \mathcal{L}(D + G_i)$ for $1 \leq i \leq r$, then for each $b = 1, \ldots, r$ we have

$$h_b = \sum_{i=1}^{r} h_i - \sum_{\substack{i=1 \\ i \neq b}}^{r} h_i \in \mathcal{L}\left(D + \sum_{\substack{i=1 \\ i \neq b}}^{r} G_i \right).$$

This implies that $\nu_P(h_b) + \nu_P(D) \geq 0$ for each place $P \in \operatorname{supp}(G_b)$. But since $h_b \in \mathcal{L}(D + G_b)$, this yields $h_b \in \mathcal{L}(D)$.

We now construct a linear code as follows. We note first that every

$$f \in \mathcal{L}\left(D + \sum_{i=1}^{r} G_i - E \right) \subseteq \mathcal{L}\left(D + \sum_{i=1}^{r} G_i \right)$$

has a unique representation

$$f = \sum_{i=1}^{r} \sum_{j=1}^{s_i} c_{i,j} f_{i,j} + u \tag{5.11}$$

with all $c_{i,j} \in \mathbb{F}_q$ and $u \in \mathcal{L}(D)$.

Definition 5.4.1. The *NXL code* $C(G_1, \ldots, G_r; D, E)$ is defined as the image of the $\mathbb{F}_q$-linear map $\eta : \mathcal{L}(D + \sum_{i=1}^{r} G_i - E) \to \mathbb{F}_q^n$ given by

$$\eta(f) = (c_{1,1}, \ldots, c_{1,s_1}, \ldots, c_{r,1}, \ldots, c_{r,s_r})$$

for all $f \in \mathcal{L}(D + \sum_{i=1}^{r} G_i - E)$, where the $c_{i,j}$ are as in (5.11).

The following theorem provides bounds on the dimension and the minimum distance of the linear code $C(G_1, \ldots, G_r; D, E)$.

Theorem 5.4.2. Let $F/\mathbb{F}_q$ be a global function field of genus g and let $G_1, \ldots, G_r$ be positive divisors of F with positive degrees $s_1, \ldots, s_r$, respectively, and with pairwise disjoint supports. Let D be a divisor of F with $\ell(D) = \deg(G) + 1 - g$. Furthermore, let E be a positive divisor of F with $\text{supp}(E) \cap \text{supp}(G_i) = \varnothing$ for $1 \leq i \leq r$ such that $m := \deg(E - D)$ satisfies

$$1 \leq m \leq \sum_{i=1}^{r} s_i - g.$$

Then $C(G_1, \ldots, G_r; D, E)$ is a linear $[n, k, d]$ code over $\mathbb{F}_q$ with

$$n = \sum_{i=1}^{r} s_i, \quad k = \ell\left(D + \sum_{i=1}^{r} G_i - E\right) \geq n - m - g + 1, \quad d \geq d_0,$$

where d_0 is the least cardinality of a subset R of $\{1, \ldots, r\}$ for which $\sum_{i \in R} s_i \geq m$. Moreover, we have $k = n - m - g + 1$ if $n - m \geq 2g - 1$.

Proof. Take a nonzero $f \in \mathcal{L}(D + \sum_{i=1}^{r} G_i - E)$ and let w be the weight of $\eta(f)$. We have the unique representation (5.11) for f. Define

$$R = \{1 \leq i \leq r : \exists j, 1 \leq j \leq s_i, \text{ with } c_{i,j} \neq 0\}.$$

Note that $|R| \leq w$. We can then write (5.11) in the form

$$f = \sum_{i \in R} h_i + u$$

with $h_i \in \mathcal{L}(D + G_i)$ for $i \in R$ and $u \in \mathcal{L}(D)$. Therefore, $f \in$ $\mathcal{L}(D + \sum_{i \in R} G_i)$. But we also have $f \in \mathcal{L}(D + \sum_{i=1}^{r} G_i - E)$, and in view of the condition on $\mathrm{supp}(E)$ this shows that

$$ f \in \mathcal{L}\left(D + \sum_{i \in R} G_i - E \right). $$

Since $f \neq 0$, this implies

$$ \deg\left(D + \sum_{i \in R} G_i - E \right) \geq 0 $$

by Corollary 3.4.4, that is, $\sum_{i \in R} s_i \geq m$. It follows that

$$ w \geq |R| \geq d_0 > 0. $$

This shows not only the desired lower bound on d, but also that η is injective. Hence,

$$ k = \ell\left(D + \sum_{i=1}^{r} G_i - E \right) $$

and the rest follows from the Riemann-Roch theorem. □

Remark 5.4.3. A simple choice for the divisors $G_1, \ldots, G_r$ in Theorem 5.4.2 is to take distinct places $P_1, \ldots, P_r$ of F. Note that these places need not be rational, but can have arbitrary degrees. This special case was considered in the paper [101] which introduced NXL codes and is used also in the following two examples.

Example 5.4.4. Let $q = 2$, let F be the rational function field over $\mathbb{F}_2$, and put $r = 6$. With the notation in Remark 5.4.3, we choose for $P_1, \ldots, P_6$ three rational places, a place of degree 2, and two places of degree 3 of F. Let D be the zero divisor and E a place of degree 7 of F. Then Theorem 5.4.2 shows that $C(P_1, \ldots, P_6; D, E)$ is a linear $[n, k, d]$ code over $\mathbb{F}_2$ with $n = 11$, $k = 5$, and $d \geq 3$; hence, $k + d \geq 8$. By comparison, the best lower bound on $k + d$ for an AG code over $\mathbb{F}_2$ of length 11 is obtained by taking $g = 8$ (compare with the tables in [41]), and then $k + d \geq 4$ by Theorem 5.2.2.

Example 5.4.5. Let $q = 3$, let F be the rational function field over $\mathbb{F}_3$, and put $r = 13$. With the notation in Remark 5.4.3, we choose for $P_1, \ldots, P_{13}$ four rational places, three places of degree 2, and six places of degree 3 of F. Let D be the zero divisor and E a place of degree 7 of F. Then Theorem 5.4.2 shows that $C(P_1, \ldots, P_{13}; D, E)$ is a linear $[n, k, d]$ code over $\mathbb{F}_3$ with $n = 28, k = 22$, and $d \geq 3$; hence, $k + d \geq 25$. By comparison, the best lower bound on $k + d$ for an AG code over $\mathbb{F}_3$ of length 28 is obtained by taking $g = 15$ (compare with the tables in [41]), and then $k + d \geq 14$ by Theorem 5.2.2.

The second construction that we present in this section is a powerful method of combining data from a global function field with (short) linear codes in order to produce a longer linear code as the output. It was introduced by Xing, Niederreiter, and Lam [135] and leads to the family of XNL codes. Let $F/\mathbb{F}_q$ be a global function field and let $P_1, \ldots, P_r$ be distinct places of F, which can have arbitrary degrees. Let G be a divisor of F with $\text{supp}(G) \cap \{P_1, \ldots, P_r\} = \varnothing$. For each $i = 1, \ldots, r$, let C_i be a linear $[n_i, k_i, d_i]$ code over $\mathbb{F}_q$ with $k_i \geq \deg(P_i)$. This last condition guarantees that there exists an $\mathbb{F}_q$-linear map ϕ_i from the residue class field of P_i into C_i, which is injective. We put $n = \sum_{i=1}^{r} n_i$, that is, n is the sum of the lengths of the codes $C_1, \ldots, C_r$.

Definition 5.4.6. The *XNL code* $C(P_1, \ldots, P_r; G; C_1, \ldots, C_r)$ is defined as the image of the $\mathbb{F}_q$-linear map $\beta : \mathcal{L}(G) \to \mathbb{F}_q^n$ given by

$$\beta(f) = \big(\phi_1(f(P_1)), \ldots, \phi_r(f(P_r))\big) \quad \text{for all } f \in \mathcal{L}(G),$$

where on the right-hand side we use concatenation of vectors.

The following result provides information on the parameters of XNL codes.

Theorem 5.4.7. Let $F/\mathbb{F}_q$ be a global function field of genus g and let $P_1, \ldots, P_r$ be distinct places of F. For each $i = 1, \ldots, r$, let C_i be a linear $[n_i, k_i, d_i]$ code over $\mathbb{F}_q$ with $k_i \geq \deg(P_i)$. Furthermore, let G be a divisor of F with $\mathrm{supp}(G) \cap \{P_1, \ldots, P_r\} = \varnothing$ and

$$g \leq \deg(G) < \sum_{i=1}^{r} \deg(P_i).$$

Then $C(P_1, \ldots, P_r; G; C_1, \ldots, C_r)$ is a linear $[n, k, d]$ code over $\mathbb{F}_q$ with

$$n = \sum_{i=1}^{r} n_i, \quad k = \ell(G) \geq \deg(G) + 1 - g, \quad d \geq d_0,$$

where d_0 is the minimum of $\sum_{i \in \overline{M}} d_i$ taken over all subsets M of $\{1, \ldots, r\}$ for which $\sum_{i \in M} \deg(P_i) \leq \deg(G)$, with $\overline{M}$ denoting the complement of M in $\{1, \ldots, r\}$. Moreover, we have $k = \deg(G) + 1 - g$ if $\deg(G) \geq 2g - 1$.

Proof. Take a nonzero $f \in \mathcal{L}(G)$ and let w be the weight of $\beta(f)$. Define

$$M = \{1 \leq i \leq r : f(P_i) = 0\}.$$

Then we get

$$w = \sum_{i \in \overline{M}} w_i \geq \sum_{i \in \overline{M}} d_i,$$

where w_i is the weight of $\phi_i(f(P_i))$ for $1 \leq i \leq r$. Furthermore, we have $f \in \mathcal{L}(G - \sum_{i \in M} P_i)$. Since $f \neq 0$, this implies

$$\deg\left(G - \sum_{i \in M} P_i\right) \geq 0$$

by Corollary 3.4.4, that is, $\sum_{i \in M} \deg(P_i) \leq \deg(G)$. Hence, $w \geq d_0 > 0$. This shows not only the desired lower bound on d, but also that β is injective. Consequently $k = \ell(G)$, and the rest follows from the Riemann-Roch theorem. $\square$

Corollary 5.4.8. In the special case where $\deg(P_i) \geq d_i$ for $1 \leq i \leq r$, the minimum distance of the linear code $C(P_1, \ldots, P_r; G; C_1, \ldots, C_r)$ satisfies

$$d \geq \sum_{i=1}^{r} d_i - \deg(G).$$

Proof. Proceed as in the proof of Theorem 5.4.7 and note that if $\sum_{i \in M} \deg(P_i) \leq \deg(G)$, then

$$\sum_{i \in \overline{M}} d_i = \sum_{i=1}^{r} d_i - \sum_{i \in M} d_i \geq \sum_{i=1}^{r} d_i - \sum_{i \in M} \deg(P_i) \geq \sum_{i=1}^{r} d_i - \deg(G).$$

This implies $d_0 \geq \sum_{i=1}^{r} d_i - \deg(G)$.   □

Remark 5.4.9. If $P_1, \ldots, P_r$ are distinct rational places of F and for each $i = 1, \ldots, r$ we choose C_i to be the trivial linear $[1, 1, 1]$ code over $\mathbb{F}_q$ and ϕ_i the identity map on $\mathbb{F}_q$, then the construction of XNL codes reduces to that of AG codes (compare with Definition 5.2.1). Theorem 5.2.2 is then a special case of Theorem 5.4.7 and Corollary 5.4.8. Note that by the argument in the proof of Theorem 5.2.7, the condition on supp(G) in Theorem 5.4.7 may be dropped.

A systematic search for good XNL codes was carried out by Ding, Niederreiter, and Xing [27]. Many excellent examples were found that improved on previous constructions of linear codes, and some of these XNL codes are optimal in the sense that they yield a linear $[n, k, d]$ code over $\mathbb{F}_q$ such that for these values of q, n, and k there is no linear $[n, k]$ code over $\mathbb{F}_q$ with minimum distance larger than d. The following examples offer a small selection of optimal XNL codes. In these examples, we have $\deg(P_i) = d_i$ for $1 \leq i \leq r$, so that the lower bound on the minimum distance is obtained immediately from Corollary 5.4.8. Further excellent examples were found recently by Xing and Yeo [137].

Example 5.4.10. Let $q = 2$ and let $F = \mathbb{F}_2(x, y)$ be the elliptic function field defined by

$$y^2 + y = x + \frac{1}{x}.$$

Then F has four rational places and two places of degree 2. We choose $r = 6$ and let P_1, P_2, P_3, P_4 be the four rational places and P_5 and P_6 the two places of degree 2 of F. Furthermore, we take $[n_i, k_i, d_i] = [1, 1, 1]$ for $1 \leq i \leq 4$ and $[n_i, k_i, d_i] = [3, 2, 2]$ for $i = 5, 6$ (note that a linear $[3, 2, 2]$ code over $\mathbb{F}_2$ exists by Example 5.2.8). Then for $m = \deg(G) = 1, \ldots, 7$, we obtain a linear $[n, k, d]$ code over $\mathbb{F}_2$ with parameters

$$n = 10, \quad k = m, \quad d \geq 8 - m.$$

The linear codes with $m = 2, 3, 4$ and $d = 8 - m$ are optimal.

Example 5.4.11. Let $q = 2$ and let $F = \mathbb{F}_2(x, y)$ be the elliptic function field defined by

$$y^2 + y = x^3.$$

Then F has three rational places and three places of degree 2. We choose $r = 6$ and let P_1, P_2, P_3 be the three rational places and P_4, P_5, P_6 the three places of degree 2 of F. Furthermore, we take $[n_i, k_i, d_i] = [1, 1, 1]$ for $1 \leq i \leq 3$ and $[n_i, k_i, d_i] = [3, 2, 2]$ for $4 \leq i \leq 6$. Then for $m = \deg(G) = 1, \ldots, 8$, we obtain a linear $[n, k, d]$ code over $\mathbb{F}_2$ with parameters

$$n = 12, \quad k = m, \quad d \geq 9 - m.$$

The linear codes with $m = 3, 5$ and $d = 9 - m$ are optimal.

Further work on the theory of NXL and XNL codes was carried out by Dorfer and Maharaj [28], Heydtmann [53], [54], and Özbudak and Stichtenoth [103].

5.5 Function-Field Codes

The construction of AG codes can be approached from a much wider perspective. A general viewpoint is that of function-field codes, which was introduced in Niederreiter and Xing [100, Chapter 6] and studied in detail by Hachenberger, Niederreiter, and Xing [49].

In the first place, a function-field code is a special kind of subspace of a global function field $F/\mathbb{F}_q$. For a finite nonempty set $\mathcal{P}$ of places of F, we write

$$\mathcal{O}_\mathcal{P} := \bigcap_{P \in \mathcal{P}} \mathcal{O}_P, \qquad M_\mathcal{P} := \bigcap_{P \in \mathcal{P}} M_P.$$

Thus, $\mathcal{O}_\mathcal{P}$ is the intersection of the valuation rings and $M_\mathcal{P}$ the intersection of the maximal ideals of the places in $\mathcal{P}$.

Definition 5.5.1. A *function-field code* (in $F/\mathbb{F}_q$ with respect to $\mathcal{P}$) is a nonzero finite-dimensional $\mathbb{F}_q$-linear subspace V of F which satisfies the two conditions

$$V \subseteq \mathcal{O}_\mathcal{P} \quad \text{and} \quad V \cap M_\mathcal{P} = \{0\}.$$

An important family of examples of function-field codes is provided by suitable Riemann-Roch spaces.

Proposition 5.5.2. Let $F/\mathbb{F}_q$ be a global function field, let $\mathcal{P}$ be a finite nonempty set of places of F, and let G be a divisor of F with $\ell(G) \geq 1$, $\text{supp}(G) \cap \mathcal{P} = \varnothing$, and

$$\deg(G) < \sum_{P \in \mathcal{P}} \deg(P).$$

Then $\mathcal{L}(G)$ is a function-field code in $F/\mathbb{F}_q$ with respect to $\mathcal{P}$.

Proof. Recall that $\mathcal{L}(G)$ is an $\mathbb{F}_q$-linear subspace of F of finite dimension $\ell(G) \geq 1$. Thus, it remains to verify the two conditions in Definition 5.5.1. First, $\text{supp}(G) \cap \mathcal{P} = \varnothing$ implies that $\mathcal{L}(G) \subseteq \mathcal{O}_\mathcal{P}$. Second, for any $f \in \mathcal{L}(G) \cap M_\mathcal{P}$ we have $f \in \mathcal{L}(G - \sum_{P \in \mathcal{P}} P)$. Thus, if we had $f \neq 0$, then

$$\deg\left(G - \sum_{P \in \mathcal{P}} P\right) \geq 0$$

by Corollary 3.4.4, a contradiction to $\deg(G) < \sum_{P \in \mathcal{P}} \deg(P)$. Therefore, we get $\mathcal{L}(G) \cap M_\mathcal{P} = \{0\}$. $\qquad\square$

Example 5.5.3. Another family of examples of function-field codes is obtained as follows. Let $\mathcal{P} = \{P_1, \ldots, P_r\}$ be a finite nonempty set of places of the global function field $F/\mathbb{F}_q$ and choose $f_1, \ldots, f_k \in \mathcal{O}_\mathcal{P}$ such that the k vectors

$$\mathbf{b}_j := (f_j(P_1), \ldots, f_j(P_r)) \in \overline{\mathbb{F}_q}^r, \quad 1 \leq j \leq k,$$

are linearly independent over $\mathbb{F}_q$, where $\overline{\mathbb{F}_q}$ is an algebraic closure of $\mathbb{F}_q$. Now let V be the $\mathbb{F}_q$-linear subspace of F spanned by $f_1, \ldots, f_k$. It is clear that $V \subseteq \mathcal{O}_\mathcal{P}$. To show that $V \cap M_\mathcal{P} = \{0\}$, take $f \in V \cap M_\mathcal{P}$ and note that we can write

$$f = \sum_{j=1}^{k} a_j f_j$$

with $a_j \in \mathbb{F}_q$ for $1 \leq j \leq k$. Since $f \in M_\mathcal{P}$, we have

$$0 = f(P_i) = \sum_{j=1}^{k} a_j f_j(P_i) \quad \text{for } 1 \leq i \leq r.$$

This implies

$$\sum_{j=1}^{k} a_j \mathbf{b}_j = \mathbf{0},$$

and so the given linear independence property of the vectors $\mathbf{b}_1, \ldots, \mathbf{b}_k$ yields $a_j = 0$ for $1 \leq j \leq k$ and, hence, $f = 0$. Thus, V is indeed a function-field code in $F/\mathbb{F}_q$ with respect to $\mathcal{P}$. For suitable $\mathcal{P}$ and k, appropriate elements $f_1, \ldots, f_k$ can be constructed by the approximation theorem.

We note also that any nonzero $\mathbb{F}_q$-linear subspace of a function-field code in $F/\mathbb{F}_q$ with respect to $\mathcal{P}$ is again a function-field code in $F/\mathbb{F}_q$ with respect to $\mathcal{P}$.

Function-field codes are used as a tool to generate linear codes. The most powerful method generalizes the construction of XNL codes in Section 5.4. Let V be a function-field code in $F/\mathbb{F}_q$ with respect to $\mathcal{P} = \{P_1, \ldots, P_r\}$, where $P_1, \ldots, P_r$ are distinct places of F. For each $i = 1, \ldots, r$, let C_i be a linear $[n_i, k_i, d_i]$ code over $\mathbb{F}_q$ with $k_i \geq \deg(P_i)$ and let ϕ_i be an $\mathbb{F}_q$-linear map from the residue class field of P_i into C_i, which is injective. Put $n = \sum_{i=1}^{r} n_i$ and define the $\mathbb{F}_q$-linear map $\gamma : V \to \mathbb{F}_q^n$ by

$$\gamma(f) = \big(\phi_1(f(P_1)), \ldots, \phi_r(f(P_r))\big) \quad \text{for all } f \in V,$$

where on the right-hand side we use concatenation of vectors. The image of V under γ is the linear code $C_\mathcal{P}(V; C_1, \ldots, C_r)$ over $\mathbb{F}_q$.

Information on the parameters of $C_\mathcal{P}(V; C_1, \ldots, C_r)$ is provided by the following theorem. For $I \subseteq \{1, \ldots, r\}$ we write $\overline{I}$ for the complement of I in

$\{1, \ldots, r\}$ and $\mathcal{P}(\bar{I}) := \{P_i : i \in \bar{I}\} \subseteq \mathcal{P}$. We define

$$d_0 = \min_I \sum_{i \in I} d_i, \qquad (5.12)$$

where the minimum is extended over all $I \subseteq \{1, \ldots, r\}$ for which $V \cap \mathsf{M}_{\mathcal{P}(\bar{I})} \neq \{0\}$. The last condition is always assumed to be satisfied for $I = \{1, \ldots, r\}$. The condition $V \cap \mathsf{M}_{\mathcal{P}} = \{0\}$ implies that $d_0 \geq 1$.

Theorem 5.5.4. The code $C_{\mathcal{P}}(V; C_1, \ldots, C_r)$ constructed above is a linear $[n, k, d]$ code over $\mathbb{F}_q$ with

$$n = \sum_{i=1}^r n_i, \qquad k = \dim(V), \qquad d \geq d_0,$$

where d_0 is given by (5.12).

Proof. Take $f \in V$ with $f \neq 0$. Then $f \in \mathcal{O}_{\mathcal{P}}$ and $f \notin \mathsf{M}_{\mathcal{P}}$ by Definition 5.5.1. Therefore, the set $I = \{1 \leq i \leq r : f(P_i) \neq 0\}$ is nonempty. With w denoting the weight of $\gamma(f)$ and w_i the weight of $\phi_i(f(P_i))$ for $1 \leq i \leq r$, we get

$$w = \sum_{i \in I} w_i \geq \sum_{i \in I} d_i.$$

Now $\bar{I} = \{1 \leq i \leq r : f(P_i) = 0\}$, and so $f \in \mathsf{M}_{\mathcal{P}(\bar{I})}$. Since $f \neq 0$, this implies $V \cap \mathsf{M}_{\mathcal{P}(\bar{I})} \neq \{0\}$. Thus, the definition of d_0 yields $w \geq d_0 \geq 1$. This shows $d \geq d_0$ and also that γ is injective; hence, $k = \dim(V)$. $\square$

For $f \in V$, the *block weight* of f (with respect to $\mathcal{P} = \{P_1, \ldots, P_r\}$) is defined to be

$$\vartheta_{\mathcal{P}}(f) = |\{1 \leq i \leq r : f(P_i) \neq 0\}|.$$

The number

$$\vartheta_{\mathcal{P}}(V) = \min \{\vartheta_{\mathcal{P}}(f) : f \in V, f \neq 0\}$$

is called the *minimum block weight* of V (with respect to $\mathcal{P}$).

Corollary 5.5.5. If the notation is arranged in such a way that $d_1 \leq d_2 \leq \cdots \leq d_r$, then the minimum distance d of the code $C_{\mathcal{P}}(V; C_1, \ldots, C_r)$ satisfies

$$d \geq \sum_{i=1}^{\vartheta_{\mathcal{P}}(V)} d_i,$$

where $\vartheta_{\mathcal{P}}(V)$ is the minimum block weight of V.

Proof. In view of Theorem 5.5.4, it suffices to show that

$$d_0 \geq \sum_{i=1}^{\vartheta_{\mathcal{P}}(V)} d_i.$$

Let $I \subseteq \{1, \ldots, r\}$ be a set for which the minimum in (5.12) is attained. Then there exists a nonzero $f \in V$ with $f \in M_{\mathcal{P}(\bar{I})}$, that is, $f(P_j) = 0$ for all $j \in \bar{I}$. Because of the minimality property of I, we have $f(P_i) \neq 0$ for all $i \in I$. Thus,

$$|I| = \vartheta_{\mathcal{P}}(f) \geq \vartheta_{\mathcal{P}}(V),$$

and so

$$d_0 = \sum_{i \in I} d_i \geq \sum_{i=1}^{\vartheta_{\mathcal{P}}(V)} d_i,$$

which is the desired inequality. $\qquad\qquad\qquad\qquad\qquad\qquad$ $\square$

Remark 5.5.6. Let $\mathcal{P} = \{P_1, \ldots, P_r\}$ be a finite nonempty set of places of $F/\mathbb{F}_q$ and let G be a divisor of F with $\ell(G) \geq 1$, $\text{supp}(G) \cap \mathcal{P} = \varnothing$, and

$$\deg(G) < \sum_{i=1}^{r} \deg(P_i).$$

Then, according to Proposition 5.5.2, $\mathcal{L}(G)$ is a function-field code in $F/\mathbb{F}_q$ with respect to $\mathcal{P}$. Any code $C_{\mathcal{P}}(\mathcal{L}(G); C_1, \ldots, C_r)$ is then an XNL code (compare with Definition 5.4.6). The values for the length and the dimension in Theorems 5.4.7 and 5.5.4 agree. Let d_0 be as in (5.12) and let ϑ_0 be the lower bound on d in Theorem 5.4.7, but with a change of notation, that is,

$$\vartheta_0 := \min_I \sum_{i \in I} d_i,$$

where the minimum is extended over all $I \subseteq \{1, \ldots, r\}$ with $\sum_{i \in \bar{I}} \deg(P_i) \leq \deg(G)$. It is easily seen that $d_0 \geq \vartheta_0$. To this end, let $I \subseteq \{1, \ldots, r\}$ be such that $\mathcal{L}(G) \cap M_{\mathcal{P}(\bar{I})} \neq \{0\}$. Then there exists a nonzero $f \in \mathcal{L}(G)$ with $v_{P_i}(f) \geq 1$ for all $i \in \bar{I}$. Then

$$f \in \mathcal{L}\left(G - \sum_{i \in \bar{I}} P_i\right),$$

and since $f \neq 0$, this implies

$$\sum_{i \in \bar{I}} \deg(P_i) \leq \deg(G)$$

by Corollary 3.4.4. The inequality $d_0 \geq \vartheta_0$ is now obvious from the definitions.

The codes $C_{\mathcal{P}}(V; C_1, \ldots, C_r)$ derived from function-field codes form a universal family of linear codes, in the sense that any linear code is obtained by this construction. In fact, as the following theorem shows, a special family of these codes suffices to represent any linear code. A stronger result, according to which an even smaller family of codes can represent any linear code, was proved by Pellikaan, Shen, and van Wee [104].

Theorem 5.5.7. Let C be an arbitrary linear $[n, k]$ code over $\mathbb{F}_q$. Then there exist a global function field $F/\mathbb{F}_q$, a set $\mathcal{P} = \{P_1, \ldots, P_n\}$ of n distinct rational places of F, and a k-dimensional function-field code V in $F/\mathbb{F}_q$ with respect to $\mathcal{P}$ such that C is equal to the code $C_{\mathcal{P}}(V; C_1, \ldots, C_n)$ with C_i being the linear $[1, 1, 1]$ code over $\mathbb{F}_q$ for $i = 1, \ldots, n$.

Proof. Since $A(q) > 0$ by Remark 4.3.8, there exists a global function field $F/\mathbb{F}_q$ possessing n distinct rational places $P_1, \ldots, P_n$. Let

$\mathbf{c}_1, \ldots, \mathbf{c}_k$ form a basis of C and put

$$\mathbf{c}_j = (c_{j,1}, \ldots, c_{j,n}) \in \mathbb{F}_q^n \quad \text{for } 1 \le j \le k.$$

By the approximation theorem (see Theorem 1.5.18), for each $j = 1, \ldots, k$ we can find an $f_j \in F$ such that

$$\nu_{P_i}(f_j - c_{j,i}) \ge 1 \quad \text{for } 1 \le i \le n.$$

Let V be the $\mathbb{F}_q$-linear subspace of F spanned by $f_1, \ldots, f_k$. From the construction of the f_j we infer $V \subseteq \mathcal{O}_\mathcal{P}$ with $\mathcal{P} = \{P_1, \ldots, P_n\}$ and also

$$f_j(P_i) = c_{j,i} \quad \text{for } 1 \le j \le k,\ 1 \le i \le n. \tag{5.13}$$

Now let $\gamma : V \to \mathbb{F}_q^n$ be the $\mathbb{F}_q$-linear map defined by

$$\gamma(f) = \big(f(P_1), \ldots, f(P_n)\big) \quad \text{for all } f \in V.$$

Then (5.13) yields

$$\gamma(f_j) = \mathbf{c}_j \quad \text{for } 1 \le j \le k.$$

Since $\mathbf{c}_1, \ldots, \mathbf{c}_k$ are linearly independent over $\mathbb{F}_q$, this implies that $f_1, \ldots, f_k$ are linearly independent over $\mathbb{F}_q$ and also $V \cap M_\mathcal{P} = \{0\}$. Thus, V is a k-dimensional function-field code in $F/\mathbb{F}_q$ with respect to $\mathcal{P}$ and it is immediate that $C = C_\mathcal{P}(V; C_1, \ldots, C_n)$. $\qquad\square$

5.6 Applications of Character Sums

The character sums discussed in Section 4.4 are useful tools for obtaining information on codes, for instance, bounds on parameters and results on the weight distribution of codes. There are many examples of such applications, but we confine the exposition to some typical instances. A more systematic coverage of the applications of character sums to coding theory can be found in Honkala and Tietäväinen [57].

Let $m \ge 2$ be an integer and let α be a primitive element of the finite field $\mathbb{F}_{2^m}$. If N is a proper positive divisor of $2^m - 1$, then $\beta = \alpha^N$ is a primitive nth root of unity in $\mathbb{F}_{2^m}$, where $n = (2^m - 1)/N$. Let t be an integer with $1 \leq t \leq (n-1)/2$. Let H be the $t \times n$ matrix over $\mathbb{F}_{2^m}$ with columns $\mathbf{h}_0, \mathbf{h}_1, \ldots, \mathbf{h}_{n-1}$, where

$$\mathbf{h}_j = (\beta^j, \beta^{3j}, \ldots, \beta^{(2t-1)j})^{\mathrm{T}} \qquad \text{for } j = 0, 1, \ldots, n - 1.$$

Then

$$C := \{\mathbf{x} \in \mathbb{F}_2^n : H\mathbf{x}^{\mathrm{T}} = \mathbf{0}\} \tag{5.14}$$

is a linear code over $\mathbb{F}_2$ of length n. In fact, C is a so-called *narrow-sense BCH code* of designed distance $2t + 1$ (see [72, Section 9.2] and [75, Chapter 9]). This means, in particular, that the minimum distance $d(C)$ satisfies $d(C) \geq 2t + 1$. With the help of character sums, the true value of $d(C)$ can be determined in many cases.

Theorem 5.6.1. Let $m \geq 2$ be an integer, let N be a proper positive divisor of $2^m - 1$, and put $n = (2^m - 1)/N$. Let t be an integer with $1 \leq t \leq (n-1)/2$. Then the minimum distance $d(C)$ of the narrow-sense BCH code C over $\mathbb{F}_2$ of length n and designed distance $2t + 1$ satisfies $d(C) = 2t + 1$ whenever $2^m \geq (2tN)^{4t+2}$.

Proof. If suffices to show that $d(C) \leq 2t + 1$ under the given hypotheses. This will follow if we can prove that there exist some (not necessarily distinct) $\mathbf{y}_1, \ldots, \mathbf{y}_{2t+1} \in \{\mathbf{h}_0, \mathbf{h}_1, \ldots, \mathbf{h}_{n-1}\}$ such that

$$\mathbf{y}_1 + \cdots + \mathbf{y}_{2t+1} = \mathbf{0}, \tag{5.15}$$

for then, after eliminating pairs of identical $\mathbf{y}_j$, we are still left with a nontrivial linear dependence relation over $\mathbb{F}_2$ between distinct vectors among $\mathbf{y}_1, \ldots, \mathbf{y}_{2t+1}$, which translates into the existence of a codeword $\mathbf{x} \in C$ with $1 \leq w(\mathbf{x}) \leq 2t + 1$.

We write $q = 2^m$ and note that each $\mathbf{y}_j$, being a column vector of H, has the form

$$\mathbf{y}_j = (c_j^N, c_j^{3N}, \ldots, c_j^{(2t-1)N})^{\mathrm{T}} \qquad \text{for some } c_j \in \mathbb{F}_q^*.$$

Thus, studying the solutions of (5.15) is equivalent to studying the solutions $(c_1, \ldots, c_{2t+1}) \in (\mathbb{F}_q^*)^{2t+1}$ of

$$\mathbf{f}(c_1) + \cdots + \mathbf{f}(c_{2t+1}) = \mathbf{0}, \tag{5.16}$$

where $\mathbf{f}(c) = (c^N, c^{3N}, \ldots, c^{(2t-1)N}) \in \mathbb{F}_q^t$ for all $c \in \mathbb{F}_q$.

Let S denote the number of solutions of (5.16). Let χ be a nontrivial additive character of $\mathbb{F}_q$ and let $\cdot$ denote the standard inner product in $\mathbb{F}_q^t$. Then

$$q^t S = \sum_{c_1,\dots,c_{2t+1} \in \mathbb{F}_q^*} \sum_{\mathbf{b} \in \mathbb{F}_q^t} \chi\left(\mathbf{b} \cdot (\mathbf{f}(c_1) + \cdots + \mathbf{f}(c_{2t+1}))\right)$$

$$= \sum_{\mathbf{b} \in \mathbb{F}_q^t} \prod_{j=1}^{2t+1} \left(\sum_{c_j \in \mathbb{F}_q^*} \chi(\mathbf{b} \cdot \mathbf{f}(c_j)) \right)$$

$$= (q-1)^{2t+1} + \sum_{\substack{\mathbf{b} \in \mathbb{F}_q^t \\ \mathbf{b} \neq \mathbf{0}}} \left(\sum_{c \in \mathbb{F}_q^*} \chi(\mathbf{b} \cdot \mathbf{f}(c)) \right)^{2t+1}.$$

Now consider the innermost sum. Note that $\mathbf{b} \cdot \mathbf{f}(c)$ with $\mathbf{b} \neq \mathbf{0}$ is a nonconstant polynomial in c of odd degree at most $(2t-1)N$. Therefore, by Theorem 4.4.5,

$$\left| \sum_{c \in \mathbb{F}_q^*} \chi(\mathbf{b} \cdot \mathbf{f}(c)) \right| \leq \left| \sum_{c \in \mathbb{F}_q} \chi(\mathbf{b} \cdot \mathbf{f}(c)) \right| + 1$$

$$\leq ((2t-1)N - 1)q^{1/2} + 1 < (2t-1)Nq^{1/2}.$$

Using $2t < n < q$, we deduce that

$$\left| \sum_{c \in \mathbb{F}_q^*} \chi(\mathbf{b} \cdot \mathbf{f}(c)) \right| < \frac{2t-1}{2t} 2t N q^{1/2} < \frac{q-1}{q} 2t N q^{1/2},$$

and so

$$q^t S > (q-1)^{2t+1} - q^t \left(\frac{(q-1)2t N}{q^{1/2}} \right)^{2t+1}$$

$$= (q-1)^{2t+1} \left(1 - (2t N)^{2t+1} q^{-1/2} \right) \geq 0,$$

since $q \geq (2t N)^{4t+2}$ by hypothesis. Hence, $S > 0$, which completes the proof. $\qquad\qquad\sqcap$

Another illustration of the method of character sums in coding theory relates to bounds on the weights of codewords. For an arbitrary prime power q, a positive integer n, and an $\mathbb{F}_q$-linear subspace L of $\mathbb{F}_q^n$, we introduce the dual

$$L^\perp := \{\mathbf{c} \in \mathbb{F}_q^n : \mathbf{c} \cdot \mathbf{x} = 0 \quad \text{for all } \mathbf{x} \in L\},$$

where $\cdot$ denotes the standard inner product in $\mathbb{F}_q^n$. Elementary linear algebra shows that $(L^\perp)^\perp = L$.

We consider, first of all, the narrow-sense BCH code C over $\mathbb{F}_2$ of length $n = 2^m - 1$ and designed distance $2t + 1$. In other words, C is obtained by the construction in (5.14) with $N = 1$, and so with $\beta = \alpha$ being a primitive element of $\mathbb{F}_{2^m}$. Note that the $t \times n$ matrix H in (5.14) has then the columns $\mathbf{h}_0, \mathbf{h}_1, \ldots, \mathbf{h}_{n-1}$ with

$$\mathbf{h}_j = (\alpha^j, \alpha^{3j}, \ldots, \alpha^{(2t-1)j})^{\mathrm{T}} \quad \text{for } j = 0, 1, \ldots, n - 1.$$

We are actually interested in the dual $C^\perp$ of C. In order to determine $C^\perp$ explicitly, we write $q = 2^m$ and use the trace map $\mathrm{Tr} : \mathbb{F}_q \to \mathbb{F}_2$. For any polynomial $f \in \mathbb{F}_q[x]$, we put

$$\mathbf{a}(f) = (a_0(f), a_1(f), \ldots, a_{n-1}(f)) \in \mathbb{F}_2^n$$

with $a_j(f) = \mathrm{Tr}(f(\alpha^j))$ for $0 \leq j \leq n - 1$.

Lemma 5.6.2. Let C be the narrow-sense BCH code over $\mathbb{F}_2$ of length $n = 2^m - 1$ and designed distance $2t + 1$, with integers $m \geq 2$ and $1 \leq t \leq (n - 1)/2$. Then the dual $C^\perp$ of C is given by

$$C^\perp = \{\mathbf{a}(f) : f \in \mathcal{F}_{q,t}\},$$

where $\mathcal{F}_{q,t}$ is the set of all polynomials of the form $f(x) = \sum_{i=1}^{t} \gamma_i x^{2i-1} \in \mathbb{F}_q[x]$ with coefficients $\gamma_1, \ldots, \gamma_t \in \mathbb{F}_q$ and where $q = 2^m$.

Proof. Let $\mathbf{c} = (c_0, c_1, \ldots, c_{n-1}) \in \mathbb{F}_2^n$ and introduce the polynomial

$$c(x) = \sum_{j=0}^{n-1} c_j x^j \in \mathbb{F}_2[x].$$

Then for the inner product of $\mathbf{c}$ and $\mathbf{a}(f) \in A := \{\mathbf{a}(f) : f \in \mathcal{F}_{q,t}\}$ we get

$$\mathbf{c} \cdot \mathbf{a}(f) = \sum_{j=0}^{n-1} c_j \mathrm{Tr}(f(\alpha^j)) = \sum_{j=0}^{n-1} c_j \mathrm{Tr}\left(\sum_{i=1}^{t} \gamma_i \alpha^{j(2i-1)}\right)$$

$$= \sum_{i=1}^{t} \mathrm{Tr}\left(\gamma_i \sum_{j=0}^{n-1} c_j \alpha^{j(2i-1)}\right) = \sum_{i=1}^{t} \mathrm{Tr}(\gamma_i c(\alpha^{2i-1})).$$

If now $\mathbf{c} \in C$, then $c(\alpha^{2i-1}) = 0$ for $1 \le i \le t$ by the definition of C, and so $\mathbf{c} \cdot \mathbf{a}(f) = 0$ for all $\mathbf{a}(f) \in A$. Hence, $C \subseteq A^{\perp}$. Conversely, let $\mathbf{c} \in A^{\perp}$, that is, $\mathbf{c} \cdot \mathbf{a}(f) = 0$ for all $\mathbf{a}(f) \in A$. For each $i = 1, \ldots, t$, choose $f(x) = \gamma_i x^{2i-1} \in \mathcal{F}_{q,t}$, then $\mathrm{Tr}(\gamma_i c(\alpha^{2i-1})) = 0$ for all $\gamma_i \in \mathbb{F}_q$, and so $c(\alpha^{2i-1}) = 0$. This shows that $A^{\perp} \subseteq C$. This implies $C = A^{\perp}$, and so $C^{\perp} = A$. $\qquad\square$

By means of character sums, we can now bound the weights of codewords of the dual $C^{\perp}$ in Lemma 5.6.2.

Theorem 5.6.3. Let $C^{\perp}$ be the dual of the narrow-sense BCH code over $\mathbb{F}_2$ of length $n = 2^m - 1$ and designed distance $2t + 1$, with integers $m \ge 2$ and $1 \le t \le (n-1)/2$. Then the weight $w(\mathbf{x})$ of any nonzero codeword $\mathbf{x} \in C^{\perp}$ satisfies

$$2^{m-1} - (t-1)2^{m/2} \le w(\mathbf{x}) \le 2^{m-1} + (t-1)2^{m/2}.$$

Proof. Put again $q = 2^m$. Let χ be the nontrivial additive character of $\mathbb{F}_q$ given by

$$\chi(\gamma) = (-1)^{\mathrm{Tr}(\gamma)} \quad \text{for all } \gamma \in \mathbb{F}_q.$$

Note that χ attains only the values 1 and -1. Thus, for any polynomial $f \in \mathbb{F}_q[x]$ we get

$$\sum_{j=0}^{n-1} \chi(f(\alpha^j)) = |\{0 \le j \le n-1 : \mathrm{Tr}(f(\alpha^j)) = 0\}|$$

$$-|\{0 \le j \le n-1 : \mathrm{Tr}(f(\alpha^j)) = 1\}|$$

$$= (n - w(\mathbf{a}(f))) - w(\mathbf{a}(f)) = n - 2w(\mathbf{a}(f)).$$

Therefore,

$$w(\mathbf{a}(f)) = \frac{1}{2}\left(n - \sum_{j=0}^{n-1} \chi(f(\alpha^j)) \right). \tag{5.17}$$

If now $\mathbf{x} \in C^\perp$ with $\mathbf{x} \ne \mathbf{0}$, then by Lemma 5.6.2 we have $\mathbf{x} = \mathbf{a}(f)$ for some nonzero $f \in \mathcal{F}_{q,t}$. Then

$$\sum_{j=0}^{n-1} \chi(f(\alpha^j)) = \sum_{\gamma \in \mathbb{F}_q} \chi(f(\gamma)) - 1.$$

Hence, Theorem 4.4.5 yields

$$-(2t-2)q^{1/2} - 1 \le \sum_{j=0}^{n-1} \chi(f(\alpha^j)) \le (2t-2)q^{1/2} - 1.$$

By combining this with (5.17), we arrive at the desired result. □

Example 5.6.4. If $t = 1$ in Theorem 5.6.3, then every nonzero codeword of $C^\perp$ has the same weight 2^{m-1}. For this value of t, the code $C^\perp$ is called a *binary simplex code*. It is a linear $[2^m - 1, m, 2^{m-1}]$ code over $\mathbb{F}_2$ and has the remarkable property that the distance between any two distinct codewords of $C^\perp$ is 2^{m-1}.

5.7 Digital Nets

In this section, we present some applications of global function fields to a subject that is related to linear codes, namely the theory of digital nets. This theory was initially developed by Niederreiter [87] and is significant for applications to quasi-Monte Carlo methods in scientific computing. It would lead too far to discuss these applications here, so we refer the interested reader to the book [88] and the survey article [90] for the links to quasi-Monte Carlo methods.

The main aim of the theory of digital nets is to provide sets of points with a highly uniform distribution in a unit cube of a Euclidean space. For an integer $s \ge 1$, let $I^s = [0, 1)^s$ denote the half-open s-dimensional unit cube in $\mathbb{R}^s$.

Let q be a prime power and let $m \geq 1$ be an integer. A digital net (over $\mathbb{F}_q$) consists of q^m points in I^s that are obtained by the following construction. Choose $m \times m$ matrices $C^{(1)}, \ldots, C^{(s)}$ over $\mathbb{F}_q$, that is, one matrix for each of the s coordinate directions of points in I^s. For a fixed column vector $\mathbf{a} \in \mathbb{F}_q^m$ of length m, we compute the matrix-vector products

$$C^{(i)}\mathbf{a} \in \mathbb{F}_q^m \quad \text{for } 1 \leq i \leq s.$$

Next we define the map $T_m : \mathbb{F}_q^m \to [0, 1)$ by

$$T_m(\mathbf{h}) = \sum_{j=1}^{m} \psi(h_j)q^{-j}$$

for $\mathbf{h} = (h_1, \ldots, h_m)^{\mathrm{T}} \in \mathbb{F}_q^m$, where $\psi : \mathbb{F}_q \to \{0, 1, \ldots, q - 1\}$ is a fixed bijection from $\mathbb{F}_q$ onto the set $\{0, 1, \ldots, q - 1\}$ of q-adic digits. Then we put

$$p^{(i)}(\mathbf{a}) = T_m\left(C^{(i)}\mathbf{a}\right) \in [0, 1) \quad \text{for } 1 \leq i \leq s.$$

In this way we get the point

$$\mathbf{p}(\mathbf{a}) = \left(p^{(1)}(\mathbf{a}), \ldots, p^{(s)}(\mathbf{a})\right) \in I^s. \tag{5.18}$$

By letting $\mathbf{a}$ range over all q^m possibilities in $\mathbb{F}_q^m$, we arrive at the desired q^m points in I^s.

For these q^m points, the type of distribution behavior we are interested in depends only on the choice of the matrices $C^{(1)}, \ldots, C^{(s)}$, which are called the *generating matrices* of the digital net. The following definition is relevant here.

Definition 5.7.1. Let k, m, and s be positive integers and let d be an integer with $0 \leq d \leq \min(k, ms)$. The system $\{\mathbf{c}_j^{(i)} \in \mathbb{F}_q^k : 1 \leq j \leq m,\ 1 \leq i \leq s\}$ of vectors is called a (d, k, m, s)-*system over* $\mathbb{F}_q$ if for any integers $d_1, \ldots, d_s$ with $0 \leq d_i \leq m$ for $1 \leq i \leq s$ and $\sum_{i=1}^{s} d_i = d$ the system $\{\mathbf{c}_j^{(i)} \in \mathbb{F}_q^k : 1 \leq j \leq d_i,\ 1 \leq i \leq s\}$ is linearly independent over $\mathbb{F}_q$ (the empty system is considered linearly independent). If $k = m$, then we speak of a (d, m, s)-*system over* $\mathbb{F}_q$.

For $1 \leq i \leq s$, let $\mathbf{c}_j^{(i)}$, $1 \leq j \leq m$, denote the row vectors of the $m \times m$ matrix $C^{(i)}$. We can now define the parameters of a digital net.

Definition 5.7.2. The point set consisting of the q^m points $\mathbf{p}(\mathbf{a})$ in (5.18) with $\mathbf{a}$ running through $\mathbb{F}_q^m$ is called a *digital* (t, m, s)-*net over* $\mathbb{F}_q$ if the system $\{\mathbf{c}_j^{(i)} \in \mathbb{F}_q^m : 1 \leq j \leq m, \ 1 \leq i \leq s\}$ of row vectors of the generating matrices forms a (d, m, s)-system over $\mathbb{F}_q$ with $d = m - t$. The integer t is called the *quality parameter* of the digital net.

To emphasize, the third parameter s of a digital net indicates the dimension in which the points of the digital net live, whereas the second parameter m provides information on the number of points in the digital net (which is q^m). For the first parameter t we have $0 \leq t \leq m$. Note that Definition 5.7.2 is always satisfied for $t = m$. The interesting question is whether we can choose the generating matrices of the digital net in such a way that we obtain nontrivial values $t < m$ (or even rather small values) for the quality parameter.

Remark 5.7.3. The definition of a digital (t, m, s)-net over $\mathbb{F}_q$ can be translated into an explicit distribution property of the points of the digital net as follows. Consider any subinterval J of I^s of the form

$$J = \prod_{i=1}^{s} [e_i q^{-d_i}, (e_i + 1)q^{-d_i})$$

with $e_i, d_i \in \mathbb{Z}$, $d_i \geq 0$, $0 \leq e_i < q^{d_i}$ for $1 \leq i \leq s$, and with J having the s-dimensional volume q^{t-m}. Then for any such J there exist exactly q^t vectors $\mathbf{a} \in \mathbb{F}_q^m$ such that the corresponding point $\mathbf{p}(\mathbf{a})$ in (5.18) lies in J. The proof is elementary and can be found, for instance, in [88, Section 4.3]. From this viewpoint, it is again clear that we are interested in smaller values of t since then the family of intervals J for which the above distribution property holds is larger.

Remark 5.7.4. A simple relationship between digital nets and linear codes arises if we consider the special case $m = 1$ of the concept of a (d, k, m, s)-system over $\mathbb{F}_q$ in Definition 5.7.1. Thus, suppose we are given a $(d, k, 1, s)$-system $\{\mathbf{c}^{(i)} \in \mathbb{F}_q^k : 1 \leq i \leq s\}$ over $\mathbb{F}_q$ with $k < s$. Then we obtain a linear code over $\mathbb{F}_q$ of length s, dimension $\geq s - k$, and minimum distance $\geq d + 1$ if we use $\mathbf{c}^{(1)}, \ldots, \mathbf{c}^{(s)}$ as the columns of a parity-check matrix of the linear code. This follows from Proposition 5.1.2. This observation suggests that the construction of good digital nets will probably be more difficult than the construction of good linear codes. Note that the construction of a good linear code requires us to find a list of s vectors (the columns of a parity-check matrix), that is, an $s \times 1$ array of vectors, with a suitable linear independence property, whereas in the construction of a good digital (t, m, s)-net over $\mathbb{F}_q$ we are looking for an $s \times m$ array of vectors $\mathbf{c}_j^{(i)} \in \mathbb{F}_q^m$, $1 \le i \le s$, $1 \le j \le m$, with a more general linear independence property.

Example 5.7.5. The following is an easy construction of an interesting digital net. Let q be an arbitrary prime power, let $m \ge 1$ be an integer, and let $s = 2$. Choose a basis $\{\mathbf{b}_1, \ldots, \mathbf{b}_m\}$ of $\mathbb{F}_q^m$ and define the generating matrix $C^{(1)}$ with row vectors $\mathbf{c}_j^{(1)} = \mathbf{b}_j$ for $1 \le j \le m$ and the generating matrix $C^{(2)}$ with row vectors $\mathbf{c}_j^{(2)} = \mathbf{b}_{m-j+1}$ for $1 \le j \le m$. Then it is clear that the system $\{\mathbf{c}_j^{(i)} \in \mathbb{F}_q^m : 1 \le j \le m, \; 1 \le i \le 2\}$ forms a $(d, m, 2)$-system over $\mathbb{F}_q$ with $d = m$. Thus, the corresponding point set is a digital $(0, m, 2)$-net over $\mathbb{F}_q$ by Definition 5.7.2.

Note that Example 5.7.5 yields two-dimensional digital nets with the smallest value $t = 0$ of the quality parameter. However, if q is fixed and $m \ge 2$, then $t = 0$ can be achieved only for finitely many dimensions, as the following result shows.

Proposition 5.7.6. Let q be a prime power and let $m \ge 2$ and $s \ge q+2$ be integers. Then there cannot exist a digital $(0, m, s)$-net over $\mathbb{F}_q$.

Proof. We proceed by contradiction and assume that there exists an (m, m, s)-system $C = \{\mathbf{c}_j^{(i)} \in \mathbb{F}_q^m : 1 \le j \le m, \; 1 \le i \le s\}$ over $\mathbb{F}_q$. Then $\{\mathbf{c}_1^{(1)}, \ldots, \mathbf{c}_m^{(1)}\}$ is a basis of $\mathbb{F}_q^m$. In the representation of each vector $\mathbf{c}_1^{(i)}$, $2 \le i \le s$, in terms of this basis, the coefficient of $\mathbf{c}_m^{(1)}$ must be nonzero by the definition of an (m, m, s)-system over $\mathbb{F}_q$. Thus, for each $i = 2, \ldots, s$, there exists a nonzero $a_i \in \mathbb{F}_q$ such that $a_i \mathbf{c}_1^{(i)} - \mathbf{c}_m^{(1)}$ is a linear combination of $\mathbf{c}_1^{(1)}, \ldots, \mathbf{c}_{m-1}^{(1)}$. Let $b_i \subset \mathbb{F}_q$ be the coefficient of $\mathbf{c}_{m-1}^{(1)}$ in the last representation. Since $s \ge q + 2$, two of the elements $b_2, \ldots, b_s$ must be identical, say $b_h = b_k$ with $2 \le h < k \le s$. Then by subtraction we see that $a_h \mathbf{c}_1^{(h)} - a_k \mathbf{c}_1^{(k)}$ is a linear combination of $\mathbf{c}_1^{(1)}, \ldots, \mathbf{c}_{m-2}^{(1)}$, a contradiction to C being an (m, m, s)-system over $\mathbb{F}_q$. $\square$

We discuss now another relationship between digital nets and linear codes. The relationship mentioned in Remark 5.7.4 arises from the consideration of parity-check matrices of linear codes, whereas the new relationship is based on generalizing the Hamming weight and the connection between Hamming weight and minimum distance in (5.1).

Let $m \geq 1$ and $s \geq 1$ be integers. First, we define a weight function v on $\mathbb{F}_q^m$ by putting $v(\mathbf{a}) = 0$ if $\mathbf{a} = \mathbf{0} \in \mathbb{F}_q^m$, and for $\mathbf{a} = (a_1, \ldots, a_m) \in \mathbb{F}_q^m$ with $\mathbf{a} \neq \mathbf{0}$ we set

$$v(\mathbf{a}) = \max \{ j : a_j \neq 0 \}.$$

Then we extend this definition to $\mathbb{F}_q^{ms}$ by writing a vector $\mathbf{A} \in \mathbb{F}_q^{ms}$ as the concatenation of s vectors of length m, that is,

$$\mathbf{A} = (\mathbf{a}^{(1)}, \ldots, \mathbf{a}^{(s)}) \in \mathbb{F}_q^{ms} \text{ with } \mathbf{a}^{(i)} \in \mathbb{F}_q^m \text{ for } 1 \leq i \leq s,$$

and putting

$$V_m(\mathbf{A}) = \sum_{i=1}^{s} v\left(\mathbf{a}^{(i)}\right).$$

The weight function V_m on $\mathbb{F}_q^{ms}$ was first introduced by Niederreiter [85] and later used in an equivalent form in coding theory by Rosenbloom and Tsfasman [108]. Note that in the case $m = 1$, the weight $V_m(\mathbf{A})$ reduces to the Hamming weight of the vector $\mathbf{A} \in \mathbb{F}_q^s$. If for any $m \geq 1$ we define the distance $d_m(\mathbf{A}, \mathbf{B})$ of $\mathbf{A}, \mathbf{B} \in \mathbb{F}_q^{ms}$ by $d_m(\mathbf{A}, \mathbf{B}) = V_m(\mathbf{A} - \mathbf{B})$, then $\mathbb{F}_q^{ms}$ turns into a metric space.

As in coding theory, the concept of minimum distance relative to d_m, or equivalently V_m, plays a crucial role.

Definition 5.7.7. For any nonzero $\mathbb{F}_q$-linear subspace $\mathcal{N}$ of $\mathbb{F}_q^{ms}$, we define the *minimum distance*

$$\delta_m(\mathcal{N}) = \min_{\mathbf{A} \in \mathcal{N} \backslash \{\mathbf{0}\}} V_m(\mathbf{A}).$$

The following is a generalization of the Singleton bound in Proposition 5.1.3. The Singleton bound corresponds to the case $m = 1$. As usual, we write $\dim(\mathcal{N})$ for the dimension of a finite-dimensional vector space $\mathcal{N}$ over $\mathbb{F}_q$.

Proposition 5.7.8. For any nonzero $\mathbb{F}_q$-linear subspace $\mathcal{N}$ of $\mathbb{F}_q^{ms}$, we have

$$\delta_m(\mathcal{N}) \leq ms - \dim(\mathcal{N}) + 1.$$

Proof. Put $h = \dim(\mathcal{N})$ and let $\pi : \mathcal{N} \to \mathbb{F}_q^h$ be the linear transformation, which maps $\mathbf{A} \in \mathcal{N}$ to the h-tuple of the last h coordinates of $\mathbf{A}$. If π is surjective, then there exists a nonzero $\mathbf{A}_1 \in \mathcal{N}$ with

$$\pi(\mathbf{A}_1) = (1, 0, \ldots, 0) \in \mathbb{F}_q^h.$$

Then

$$V_m(\mathbf{A}_1) \leq ms - h + 1.$$

If π is not surjective, then for any nonzero $\mathbf{A}_2$ in the kernel of π we have

$$V_m(\mathbf{A}_2) \leq ms - h.$$

In both cases we get the result of the proposition. □

Now we associate with a given s-dimensional digital net, having $m \times m$ generating matrices $C^{(1)}, \ldots, C^{(s)}$ over $\mathbb{F}_q$, an $\mathbb{F}_q$-linear subspace of $\mathbb{F}_q^{ms}$. First we set up an $m \times ms$ matrix $M = M(C^{(1)}, \ldots, C^{(s)})$ as follows: for $1 \leq j \leq m$, the jth row of M is obtained by concatenating the transposes of the jth column vectors of $C^{(1)}, \ldots, C^{(s)}$. The row space $\mathcal{M} \subseteq \mathbb{F}_q^{ms}$ of the matrix M is called the *row space* of the digital net. We note that $\dim(\mathcal{M}) \leq m$.

We define the dual $\mathcal{M}^\perp$ of the row space $\mathcal{M}$ as in Section 5.6, that is,

$$\mathcal{M}^\perp = \{\mathbf{C} \in \mathbb{F}_q^{ms} : \mathbf{C} \cdot \mathbf{X} = 0 \quad \text{for all } \mathbf{X} \in \mathcal{M}\}.$$

Then $\dim(\mathcal{M}^\perp) = ms - \dim(\mathcal{M}) \geq ms - m$. We assume that $s \geq 2$ since the case of one-dimensional digital nets is trivial. Then $\mathcal{M}^\perp$ is a nonzero $\mathbb{F}_q$-linear subspace of $\mathbb{F}_q^{ms}$. We can now establish the following basic result due to Niederreiter and Pirsic [97].

Theorem 5.7.9. Let $m \geq 1$ and $s \geq 2$ be integers and let $\mathcal{M} \subseteq \mathbb{F}_q^{ms}$ be the row space of an s-dimensional digital net with $m \times m$ generating matrices over $\mathbb{F}_q$. Then the digital net has quality parameter t if and only if the dual $\mathcal{M}^\perp$ of $\mathcal{M}$ satisfies

$$\delta_m(\mathcal{M}^\perp) \geq m - t + 1.$$

Proof. Let $C^{(1)}, \ldots, C^{(s)}$ be the generating matrices of the given digital net. For $1 \leq i \leq s$, let $\mathbf{c}_j^{(i)}$, $1 \leq j \leq m$, be the row vectors of $C^{(i)}$. In view of Definition 5.7.2, we have to show that $\{\mathbf{c}_j^{(i)} \in \mathbb{F}_q^m : 1 \leq j \leq m, 1 \leq i \leq s\}$ is a (d, m, s)-system over $\mathbb{F}_q$ if and only if $\mathcal{M}^\perp$ satisfies $\delta_m(\mathcal{M}^\perp) \geq d + 1$.

Put $M = M(C^{(1)}, \ldots, C^{(s)})$. Then for a row vector $\mathbf{A} = (\mathbf{a}^{(1)}, \ldots, \mathbf{a}^{(s)}) \in \mathbb{F}_q^{ms}$ with

$$\mathbf{a}^{(i)} = (a_1^{(i)}, \ldots, a_m^{(i)}) \in \mathbb{F}_q^m \quad \text{for } 1 \leq i \leq s$$

we have

$$\sum_{i=1}^{s} \sum_{j=1}^{m} a_j^{(i)} \mathbf{c}_j^{(i)} = \mathbf{0} \in \mathbb{F}_q^m$$

if and only if

$$M\mathbf{A}^{\mathrm{T}} = \mathbf{0} \in \mathbb{F}_q^m,$$

that is, if and only if $\mathbf{A} \in \mathcal{M}^\perp$.

Now assume that $\{\mathbf{c}_j^{(i)} \in \mathbb{F}_q^m : 1 \leq j \leq m, 1 \leq i \leq s\}$ is a (d, m, s)-system over $\mathbb{F}_q$. Consider any nonzero $\mathbf{A} \in \mathcal{M}^\perp$. Then from the above we get

$$\sum_{i=1}^{s} \sum_{j=1}^{m} a_j^{(i)} \mathbf{c}_j^{(i)} = \mathbf{0} \in \mathbb{F}_q^m.$$

Put $v(\mathbf{a}^{(i)}) = v_i$ for $1 \leq i \leq s$. Then

$$\sum_{i=1}^{s} \sum_{j=1}^{v_i} a_j^{(i)} \mathbf{c}_j^{(i)} = \mathbf{0} \in \mathbb{F}_q^m.$$

Since not all coefficients $a_j^{(i)}$ are 0, the system $\{\mathbf{c}_j^{(i)} \in \mathbb{F}_q^m : 1 \leq j \leq v_i, 1 \leq i \leq s\}$ is linearly dependent over $\mathbb{F}_q$. Thus, the definition of a

(d, m, s)-system over $\mathbb{F}_q$ implies that $\sum_{i=1}^{s} v_i \geq d + 1$. Therefore,

$$V_m(\mathbf{A}) = \sum_{i=1}^{s} v\left(\mathbf{a}^{(i)}\right) = \sum_{i=1}^{s} v_i \geq d + 1,$$

and so $\delta_m(\mathcal{M}^\perp) \geq d + 1$.

Conversely, assume that $\delta_m(\mathcal{M}^\perp) \geq d + 1$. We have to show that any system $\{\mathbf{c}_j^{(i)} \in \mathbb{F}_q^m : 1 \leq j \leq d_i, 1 \leq i \leq s\}$ with integers $0 \leq d_i \leq m$ for $1 \leq i \leq s$ and $\sum_{i=1}^{s} d_i = d$ is linearly independent over $\mathbb{F}_q$. Suppose, on the contrary, that such a system were linearly dependent over $\mathbb{F}_q$, that is, that there exist coefficients $a_j^{(i)} \in \mathbb{F}_q$, not all 0, such that

$$\sum_{i=1}^{s} \sum_{j=1}^{d_i} a_j^{(i)} \mathbf{c}_j^{(i)} = \mathbf{0} \in \mathbb{F}_q^m.$$

Define $a_j^{(i)} = 0$ for $d_i < j \leq m, \ 1 \leq i \leq s$. Then

$$\sum_{i=1}^{s} \sum_{j=1}^{m} a_j^{(i)} \mathbf{c}_j^{(i)} = \mathbf{0} \in \mathbb{F}_q^m.$$

By what we have shown earlier in the proof, this leads to a nonzero vector $\mathbf{A} \in \mathcal{M}^\perp$. Hence, $\delta_m(\mathcal{M}^\perp) \geq d + 1$ implies that $V_m(\mathbf{A}) \geq d + 1$. On the other hand, $v(\mathbf{a}^{(i)}) \leq d_i$ for $1 \leq i \leq s$ by the definition of the $a_j^{(i)}$, and so

$$V_m(\mathbf{A}) = \sum_{i=1}^{s} v\left(\mathbf{a}^{(i)}\right) \leq \sum_{i=1}^{s} d_i = d,$$

which yields the desired contradiction. $\square$

Corollary 5.7.10. Let $m \geq 1$ and $s \geq 2$ be integers. Then from any $\mathbb{F}_q$-linear subspace $\mathcal{N}$ of $\mathbb{F}_q^{ms}$ with $\dim(\mathcal{N}) \geq ms - m$ we can construct a digital (t, m, s)-net over $\mathbb{F}_q$ with $t = m - \delta_m(\mathcal{N}) + 1$.

Proof. Put $\mathcal{M} = \mathcal{N}^\perp \subseteq \mathbb{F}_q^{ms}$. Then $\dim(\mathcal{M}) \leq m$ and $\mathcal{M}$ can be viewed as the row space of an $m \times ms$ matrix $M(C^{(1)}, \ldots, C^{(s)})$ over $\mathbb{F}_q$. By

Theorem 5.7.9, the digital net with generating matrices $C^{(1)}, \ldots, C^{(s)}$ is a digital (t, m, s)-net over $\mathbb{F}_q$, where the quality parameter t is possible as long as $t \geq m - \delta_m(\mathcal{N}) + 1$. Now $1 \leq \delta_m(\mathcal{N}) \leq m + 1$ by Proposition 5.7.8, and so $m - \delta_m(\mathcal{N}) + 1$ lies in the interval $[0, m]$ and is thus a possible value of t.   □

Remark 5.7.11. There is a more general version of Theorem 5.7.9, which holds for (d, k, m, s)-systems over $\mathbb{F}_q$ (see [97]). By proceeding as in Corollary 5.7.10, this version can be used for the construction of (d, k, m, s)-systems over $\mathbb{F}_q$.

Corollary 5.7.10 is an important tool for finding analogs of constructions of good linear codes in the realm of digital nets. In the context of the theory of global function fields, a powerful construction of linear codes is that of AG codes (see Section 5.2). We present now an extension of the construction principle of AG codes to the setting of digital nets which uses Corollary 5.7.10. This construction of digital nets is a special case of a family of constructions introduced by Niederreiter and Özbudak [93].

Let $F/\mathbb{F}_q$ be a global function field. For a given dimension $s \geq 2$, we assume that $N(F) \geq s$ and let $P_1, \ldots, P_s$ be s distinct rational places of F. For each $i = 1, \ldots, s$, let $t_i \in F$ be a local parameter at P_i. Next we choose an arbitrary divisor G of F and put

$$n_i = v_{P_i}(G) \quad \text{for } 1 \leq i \leq s.$$

Consider the Riemann-Roch space $\mathcal{L}(G)$ and note that $v_{P_i}(f) \geq -n_i$ for $1 \leq i \leq s$ and any $f \in \mathcal{L}(G)$. For a given integer $m \geq 1$, there exists therefore a uniquely determined $\theta^{(i)}(f) \in \mathbb{F}_q^m$ with

$$v_{P_i}\left(f - \theta^{(i)}(f) \cdot (t_i^{-n_i+m-1}, t_i^{-n_i+m-2}, \ldots, t_i^{-n_i})\right) \geq -n_i + m.$$

In other words, the coordinates of $\theta^{(i)}(f)$ are, in descending order, the coefficients of t_i^j, $j = -n_i + m - 1, -n_i + m - 2, \ldots, -n_i$, in the local expansion of f at P_i (compare with (1.4)). Now we set up the $\mathbb{F}_q$-linear map $\theta : \mathcal{L}(G) \to \mathbb{F}_q^{ms}$ given by

$$\theta(f) = (\theta^{(1)}(f), \ldots, \theta^{(s)}(f)) \quad \text{for all } f \in \mathcal{L}(G).$$

Proposition 5.7.12. Let $\mathcal{N}$ be the image of the map θ. If $g \leq \deg(G) < ms$, then $\mathcal{N}$ is an $\mathbb{F}_q$-linear subspace of $\mathbb{F}_q^{ms}$ satisfying

$$\dim(\mathcal{N}) \geq \deg(G) + 1 - g, \quad \delta_m(\mathcal{N}) \geq ms - \deg(G),$$

where g is the genus of the global function field $F/\mathbb{F}_q$.

Proof. Take any nonzero $f \in \mathcal{L}(G)$ and put

$$w_i(f) = \min(m, v_{P_i}(f) + n_i) \quad \text{for } 1 \leq i \leq s.$$

The definition of $\theta^{(i)}(f)$ implies that

$$v(\theta^{(i)}(f)) = m - w_i(f) \quad \text{for } 1 \leq i \leq s.$$

Therefore,

$$V_m(\theta(f)) = \sum_{i=1}^{s} v(\theta^{(i)}(f)) = ms - \sum_{i=1}^{s} w_i(f).$$

For each $i = 1, \ldots, s$, we have $v_{P_i}(f) \geq w_i(f) - n_i$, and so

$$f \in \mathcal{L}\left(G - \sum_{i=1}^{s} w_i(f) P_i\right).$$

Since $f \neq 0$, we can use Corollary 3.4.4 to get

$$0 \leq \deg\left(G - \sum_{i=1}^{s} w_i(f) P_i\right) = \deg(G) - \sum_{i=1}^{s} w_i(f)$$
$$= \deg(G) + V_m(\theta(f)) - ms,$$

and so

$$V_m(\theta(f)) \geq ms - \deg(G) > 0.$$

This shows that the map θ is injective. Hence,

$$\dim(\mathcal{N}) = \ell(G) \geq \deg(G) + 1 - g$$

by Riemann's theorem. It is clear from the above that $\delta_m(\mathcal{N}) \geq ms - \deg(G)$. $\qquad \square$

Theorem 5.7.13. Let q be an arbitrary prime power. If $m \geq \max(1, g)$ and $s \geq 2$ are integers, then we can construct a digital (g, m, s)-net over $\mathbb{F}_q$ whenever there is a global function field $F/\mathbb{F}_q$ of genus g with $N(F) \geq s$.

Proof. In the construction leading to Proposition 5.7.12 we choose a divisor G of F with

$$\deg(G) = ms - m + g - 1.$$

Hence, $g \leq \deg(G) < ms$. Then the $\mathbb{F}_q$-linear subspace $\mathcal{N}$ of $\mathbb{F}_q^{ms}$ in Proposition 5.7.12 satisfies

$$\dim(\mathcal{N}) \geq ms - m, \qquad \delta_m(\mathcal{N}) \geq m - g + 1.$$

The rest follows from Corollary 5.7.10. $\qquad \square$

Example 5.7.14. Let q be an arbitrary prime power. Then for every integer $m \geq 1$ and every dimension s with $1 \leq s \leq q + 1$, we can construct a digital $(0, m, s)$-net over $\mathbb{F}_q$. For $s = 1$, it suffices to choose an $m \times m$ generating matrix $C^{(1)}$ over $\mathbb{F}_q$, which is nonsingular. For $2 \leq s \leq q + 1$, we apply Theorem 5.7.13 with F being the rational function field over $\mathbb{F}_q$. We note that the upper bound $s \leq q + 1$ on the dimension s is best possible, in the sense that if $s \geq q + 2$ and $m \geq 2$, then there cannot exist a digital $(0, m, s)$-net over $\mathbb{F}_q$ by Proposition 5.7.6.

The above construction of digital nets based on global function fields can be extended in various ways. For instance, there is a construction using arbitrary places of a global function field instead of just rational places, and the construction principle can be generalized to obtain (d, k, m, s)-systems (see [93]). In the following, we show that a method we used in

Section 5.3 to improve on the construction of AG codes in certain cases (see Proposition 5.3.3) has an analog in the context of digital nets. We recall our standard notation $h(F)$ for the divisor class number of a global function field F and $A_k(F)$ for the number of positive divisors of F of degree k.

Theorem 5.7.15. Let $s \geq 2$ be a given dimension and let $F/\mathbb{F}_q$ be a global function field of genus $g \geq 1$ with $N(F) \geq s$. If k and m are integers with $0 \leq k \leq g - 1$ and $m \geq \max(1, g - k - 1)$, then there exists a digital $(g - k - 1, m, s)$-net over $\mathbb{F}_q$ provided that

$$\binom{s + m + k - g}{s - 1} A_k(F) < h(F). \tag{5.19}$$

Proof. Since $N(F) \geq s$, we can choose s distinct rational places $P_1, \ldots, P_s$ of F. Note that

$$\binom{s + m + k - g}{s - 1} A_k(F)$$

is an upper bound on the number of divisors of F of the form $D + \sum_{i=1}^{s} u_i P_i$ with a positive divisor D of F of degree k and integers $0 \leq u_i \leq m$ for $1 \leq i \leq s$ such that

$$\sum_{i=1}^{s} u_i = ms - m + g - k - 1. \tag{5.20}$$

For any fixed degree, there are exactly $h(F)$ distinct divisor classes of F of that degree. Hence, by condition (5.19), there exists a divisor G of F with $\deg(G) = ms - m + g - 1$, which is not equivalent to any of the divisors $D + \sum_{i=1}^{s} u_i P_i$ considered above. We claim that

$$\mathcal{L}\left(G - \sum_{i=1}^{s} u_i P_i\right) = \{0\} \tag{5.21}$$

for all integers u_i with $0 \leq u_i \leq m$ for $1 \leq i \leq s$ satisfying (5.20). Suppose, on the contrary, that for some such $u_1, \ldots, u_s$ there exists a nonzero $b \in \mathcal{L}(G - \sum_{i=1}^{s} u_i P_i)$. Then

$$E := \mathrm{div}(b) + G - \sum_{i=1}^{s} u_i P_i \geq 0.$$

Thus, E is a positive divisor of F with $\deg(E) = k$ and

$$G = E + \sum_{i=1}^{s} u_i P_i - \mathrm{div}(b)$$

is equivalent to $E + \sum_{i=1}^{s} u_i P_i$, a contradiction to the choice of G. Thus, the claim (5.21) is established.

Now we consider the $\mathbb{F}_q$-linear subspace $\mathcal{N}$ of $\mathbb{F}_q^{ms}$, which is defined as in Proposition 5.7.12. As in the proof of Proposition 5.7.12, for any nonzero $f \in \mathcal{L}(G)$ we obtain

$$V_m(\theta(f)) = ms - \sum_{i=1}^{s} w_i(f)$$

and

$$f \in \mathcal{L}\left(G - \sum_{i=1}^{s} w_i(f) P_i\right).$$

Since $f \neq 0$, it follows from (5.21) that we must have

$$\sum_{i=1}^{s} w_i(f) \leq ms - m + g - k - 2,$$

hence,

$$V_m(\theta(f)) \geq m - g + k + 2 > 0.$$

This shows that the map θ is injective. Thus,

$$\dim(\mathcal{N}) = \ell(G) \geq \deg(G) + 1 - g = ms - m$$

and also $\delta_m(\mathcal{N}) \geq m - g + k + 2$. The rest follows from Corollary 5.7.10.

□

Example 5.7.16. Let F be the Hermitian function field over $\mathbb{F}_9$. Then $g = 3$, $N(F) = 28$, and $h(F) = L(F, 1) = 4096$ (see Example 4.2.6). We apply Theorem 5.7.15 with $s = 28$, $k = 0$, and $m = 5$. Note that

$$\binom{s + m + k - g}{s - 1} = \binom{30}{27} = \binom{30}{3} = 4060$$

and $A_0(F) = 1$, hence the condition (5.19) is satisfied. Thus, we obtain a digital $(2, 5, 28)$-net over $\mathbb{F}_9$. The value $t = 2$ agrees with the currently best value of the quality parameter for a digital $(t, 5, 28)$-net over $\mathbb{F}_9$.

The recent survey articles [91] and [92] discuss further constructions of digital nets and also provide references to tables and databases for parameters of digital nets.

6   Applications to Cryptography

Cryptography is concerned with the design of schemes that aim to resolve security issues such as the confidentiality, authenticity, and integrity of sensitive data. Cryptography has become a major research area, which is driven by the demands of modern information and communication technology. This chapter is not meant to survey this large area, but presents instead a sample of cryptographic constructions in which algebraic curves over finite fields, or equivalently global function fields, are involved. For systematic treatments of cryptography, we refer to the encyclopedic handbook by Menezes, van Oorschot, and Vanstone [80] and to the textbooks of Buchmann [12], Delfs and Knebl [24], Stinson [119], and van Tilborg [121]. Further applications of global function fields to cryptography can be found in the monograph of Niederreiter and Xing [100, Chapter 7] and in the survey article of Niederreiter, Wang, and Xing [98].

We set the stage in Section 6.1 by providing some background on cryptography and reviewing group-based cryptographic schemes. The important special case where the group is that of $\mathbb{F}_q$-rational points of an elliptic curve over $\mathbb{F}_q$ leads to elliptic-curve cryptosystems, which are discussed in Section 6.2. The case where the group is that of divisor classes of degree 0 of a global function field, and in particular of a hyperelliptic function field, is considered in Section 6.3. Interesting connections between coding theory and cryptography form the subject of Section 6.4. Some remarkable applications of global function fields to the construction of schemes for intellectual property protection are covered in Section 6.5. The efficient implementation of many cryptographic schemes requires fast arithmetic in finite fields. Section 6.6 shows how global function fields with many rational places can be used to speed up multiplication in large finite fields.

6.1 Background on Cryptography

One of the basic tasks of cryptography is to guarantee the confidentiality of sensitive data. Cryptosystems are schemes that are designed to achieve this goal. A *cryptosystem* transforms the original data (i.e., the *plaintext*) into encrypted data (i.e., the *ciphertext*) and allows unique recovery of the plaintext from the ciphertext by decryption. The encryption and decryption algorithms depend on the choice of parameters called *keys*, with the provision that the number of possible keys is so large that it inhibits exhaustive search by an attacker.

A fundamental distinction is made between public-key (or asymmetric) cryptosystems and symmetric cryptosystems. In a *public-key cryptosystem*, the encryption key K is public knowledge and only the decryption key K' is kept secret from unauthorized users. In a *symmetric cryptosystem*, both keys K and K' have to be kept secret since they are computationally equivalent in the sense that one can easily be obtained from the other; in many cases we even have $K = K'$. A typical example of a public-key cryptosystem is RSA (see Example 6.1.1 below) and typical examples of symmetric cryptosystems are the de facto industry standards DES and AES.

Example 6.1.1. Although it does not depend on algebraic geometry but just on elementary number theory, we sketch the design of the *RSA cryptosystem* since it is an easily comprehended example of a public-key cryptosystem. Consider two users, Alice and Bob, and suppose that Alice wants to send a message containing sensitive data to Bob via an insecure communication channel (e.g., the Internet). Bob publishes a public key consisting of the integers n and e that are obtained as follows. First, Bob chooses two distinct large primes p_1 and p_2 and computes $n = p_1 p_2$ and $\phi(n) = (p_1 - 1)(p_2 - 1)$. Next, Bob selects an integer $e \geq 2$ with $\gcd(e, \phi(n)) = 1$. The private key of Bob, which is kept secret by him, consists of the numbers p_1, p_2, and d, where d is a positive integer solving the congruence $ed \equiv 1 \pmod{\phi(n)}$. Now Alice formats her message in such a way that it is an integer m with $2 \leq m \leq n - 2$. Given the plaintext m, Alice computes $c := m^e \pmod{n}$ with $1 \leq c \leq n - 1$ by using Bob's public key and sends c as the ciphertext. Upon receiving c, Bob uses his secret key d to compute the least residue of c^d modulo n, which is easily seen to be m. The security of the RSA cryptosystem relies on the presumed computational infeasibility of inferring d from the public key n and e. To this end, current guidelines recommend that the primes p_1 and p_2 should have about 150 decimal digits.

Some of the cryptographic schemes that we present are based on finite abelian groups. Several interesting and suitable finite abelian groups arise from the material in this book; for instance, the multiplicative group of a finite field $\mathbb{F}_q$, the group of $\mathbb{F}_q$-rational points of an elliptic curve over $\mathbb{F}_q$ (see Section 3.7), and the group of divisor classes of degree 0 of a global function field (see Section 4.1).

We start the discussion of group-based cryptographic schemes by addressing a fundamental problem arising in symmetric cryptosystems, namely that of establishing a common key between two users who communicate via an insecure channel. A simple and elegant solution of this problem was proposed by Diffie and Hellman in their seminal paper [26] on public-key cryptography. We present the *Diffie-Hellman key-agreement scheme* in a somewhat more abstract form. Let A be a finite abelian group of order $|A|$, which is written multiplicatively, and let $a \in A$ be an element of large order (thus, necessarily, $|A|$ must also be large). The group A and the element a are known to all users. If the users Alice and Bob want to establish a common key, then Alice chooses a random integer h and sends the element a^h to Bob. Similarly, Bob selects a random integer k and sends the element a^k to Alice. The common key for Alice and Bob is then the element a^{hk}, which Alice computes as $(a^k)^h$ and Bob computes as $(a^h)^k$.

The eavesdropper Eve sees a^h and a^k going over the channel. Eve can obtain h and k if she can solve the *discrete logarithm problem* for the cyclic group $\langle a \rangle$ generated by $a \in A$ (or, in short, the discrete logarithm problem for A): given $b \in \langle a \rangle$, determine the unique integer r with $b = a^r$ and $0 \le r < |\langle a \rangle|$. Thus, a necessary condition for the security of the Diffie-Hellman key-agreement scheme is that the discrete logarithm problem for $\langle a \rangle$ is computationally infeasible for the overwhelming majority of elements $b \in \langle a \rangle$. It is in this sense that we use the informal phrase that the discrete logarithm problem for A is computationally infeasible. For instance, there are many large finite fields $\mathbb{F}_q$ for which the discrete logarithm problem for the multiplicative group of $\mathbb{F}_q$ is considered to be computationally infeasible. We remark in passing that the Diffie-Hellman key-agreement scheme operates in the cyclic group generated by a, and so A can be replaced by any (even a nonabelian) group as long as the cyclic group generated by $a \in A$ has a large order.

Next we introduce a group-based public-key cryptosystem, namely the *ElGamal cryptosystem*. Let A again be a multiplicatively written finite abelian group and let $a \in A$ have large order. The data A and a are public knowledge. The message m that Alice wants to send to Bob is now assumed to be an element of A. Bob chooses an integer h as his private key and publishes $b = a^h$ as his public key. Alice selects a random integer k and computes the elements $c_1 = a^k$ and $c_2 = mb^k$ of A. The pair (c_1, c_2) is sent as the ciphertext to Bob. Then Bob can recover the plaintext m by computing $c_2 c_1^{-h}$, since

$$c_2 c_1^{-h} = mb^k (a^k)^{-h} = m a^{hk} a^{-hk} = m.$$

It is clear that a necessary condition for the security of the ElGamal cryptosystem is the computational infeasibility of the discrete logarithm problem for A.

An important method of providing authentication for sensitive data proceeds by appending a *digital signature* to the data. A digital signature depends on the plaintext and on the identity of the signer, with the signer's identity being represented by a private key. Since anybody should be able to verify a digital signature, the verification procedure uses a public key of the signer. There are obvious parallels here with the way a public-key cryptosystem operates. In fact, if we have a public-key cryptosystem, which satisfies the additional condition that the decryption algorithm can be applied to every plaintext, then it is easy to turn this public-key cryptosystem into a digital signature scheme.

Example 6.1.2. In the RSA cryptosystem (see Example 6.1.1), it is obvious that the decryption algorithm (which consists of computing the least residue of c^d modulo n) can be applied to every plaintext. We obtain the *RSA signature scheme* by using the RSA decryption algorithm for signing and the RSA encryption algorithm for verification. In detail, let p_1, p_2, n, e, and d be as in Example 6.1.1. Bob's public key consists again of n and e and his private key consists of p_1, p_2, and d. The plaintext is again an integer m with $2 \leq m \leq n - 2$. Bob signs the plaintext m by computing the digital signature $s := m^d \pmod{n}$ with $1 \leq s \leq n - 1$ and then sends the pair (m, s) to Alice. Upon receiving (m, s), Alice verifies the digital signature s by computing the least residue of s^e modulo n and checking whether it agrees with m. Clearly, the security considerations for the RSA signature scheme are the same as those for the RSA cryptosystem.

The above condition that the decryption algorithm of the public-key cryptosystem can be applied to every plaintext is not satisfied for the ElGamal cryptosystem, for the simple reason that the plaintexts (which are elements of A) and the ciphertexts (which are pairs of elements of A) have different formats. However, by an easy modification we can obtain the *ElGamal signature scheme*. Let A and $a \in A$ be as in the ElGamal cryptosystem and put $t = |\langle a \rangle|$. We also need a publicly known bijection η from $\langle a \rangle$ to $\{0, 1, \ldots, t - 1\}$. As in the ElGamal cryptosystem, Bob has the private key $h \in \mathbb{Z}$ and the public key $b = a^h$. Bob's plaintext is an integer $m \in \{0, 1, \ldots, t - 1\}$. To sign m, Bob chooses a random integer k with $\gcd(k, t) = 1$ and computes $s_1 = a^k$ as well as a solution s_2 of the congruence

$$ks_2 \equiv m - h\eta(s_1) \pmod{t}.$$

Bob sends the triple (m, s_1, s_2) to Alice. Upon receiving (m, s_1, s_2), Alice verifies the digital signature (s_1, s_2) by computing $b^{\eta(s_1)}s_1^{s_2}$ and checking whether it agrees with a^m. Note that

$$b^{\eta(s_1)}s_1^{s_2} = a^{h\eta(s_1)}a^{ks_2} = a^{h\eta(s_1)+ks_2} = a^m,$$

and so the verification procedure does indeed make sense. The security considerations for the ElGamal signature scheme are the same as those for the ElGamal cryptosystem.

A related scheme is the *Nyberg-Rueppel signature scheme*. Let A and $a \in A$ be as above and put $t = |\langle a \rangle|$ and $u = |A|$. Let κ be a publicly known bijection from A to $\{0, 1, \ldots, u - 1\}$. Bob has the private key $h \in \mathbb{Z}$ and the public key $b = a^h$. Bob's plaintext is now an element $m \in A$. To sign m, Bob chooses a random integer k with $\gcd(k, t) = 1$ and computes $s_1 = a^k m$ as well as a solution s_2 of the congruence

$$ks_2 \equiv 1 - h\kappa(s_1) \pmod{t}.$$

Bob sends the triple (m, s_1, s_2) to Alice. Upon receiving (m, s_1, s_2), Alice verifies the digital signature (s_1, s_2) by computing $b^{\kappa(s_1)}s_1^{s_2}$ and checking whether it agrees with am^{s_2}. Note that

$$b^{\kappa(s_1)}s_1^{s_2} = a^{h\kappa(s_1)}a^{ks_2}m^{s_2} = a^{h\kappa(s_1)+ks_2}m^{s_2} = am^{s_2},$$

and so the verification procedure does indeed make sense. The security considerations for the Nyberg-Rueppel signature scheme are similar to those for the ElGamal cryptosystem.

6.2 Elliptic-Curve Cryptosystems

From the previous section, we know that as soon as we have a group A with an element a of large finite order, we obtain a discrete logarithm problem and it can potentially be used to design cryptographic schemes such as the Diffie-Hellman key-agreement scheme, the ElGamal public-key cryptosystem, and the ElGamal digital signature scheme. In order to arrive at secure and practicable schemes, we require the following additional conditions on this group and the element a:

(i) the discrete logarithm problem for A (or, more precisely, the discrete logarithm problem for the cyclic group $\langle a \rangle$) must be computationally infeasible;

(ii) the computations involved in the cryptographic schemes can be implemented easily.

If A is taken to be the multiplicative group of a finite field, we get a well-known discrete logarithm problem (see [12, Chapter 9]). Some other groups from our book that lead to potentially hard discrete logarithm problems include the groups of divisor classes of degree 0 of global function fields. In this section, we focus on a special case that has received a lot of attention in the literature, namely the group of divisor classes of degree 0 of an elliptic function field over a finite field. This case was first proposed for the use in cryptography by Koblitz [62] and Miller [82].

One convenient property of the group of divisor classes of degree 0 of an elliptic function field over $\mathbb{F}_q$ is that it is isomorphic to the group of $\mathbb{F}_q$-rational points of the corresponding elliptic curve over $\mathbb{F}_q$ (see Theorem 3.7.3 and the remarks following it). In the case of elliptic function fields, we prefer to consider the group of $\mathbb{F}_q$-rational points of the elliptic curve instead of the group of divisor classes of degree 0. One advantage of this viewpoint is that we can explicitly compute the group addition in the group of $\mathbb{F}_q$-rational points, that is, the sum of two $\mathbb{F}_q$-rational points can be expressed explicitly in terms of their coordinates (see Corollary 3.7.9).

Let $(\mathcal{E}, O)$ be an elliptic curve over $\mathbb{F}_q$ defined by a nonsingular Weierstrass equation (3.8). Let $\mathcal{E}(\mathbb{F}_q)$ denote the set of $\mathbb{F}_q$-rational points of $\mathcal{E}$. It is a finite abelian group under the operation $\oplus$ introduced in Section 3.7. By the Hasse-Weil bound (see Theorem 4.2.4), we have

$$q + 1 - 2q^{1/2} \leq |\mathcal{E}(\mathbb{F}_q)| \leq q + 1 + 2q^{1/2}.$$

In order to get a large order of $\mathcal{E}(\mathbb{F}_q)$, we have to choose a large value of q. In the framework of the ElGamal cryptosystem for an abstract abelian group described in Section 6.1, we can design a public-key cryptosystem based on elliptic curves. We describe the ElGamal elliptic-curve cryptosystem as follows.

Choose a point $P \in \mathcal{E}(\mathbb{F}_q)$ such that it has a large order in the group $\mathcal{E}(\mathbb{F}_q)$. The curve $(\mathcal{E}, O)$ and the point P are public knowledge. The message that Alice wants to send to Bob is assumed to be a point $M \in \mathcal{E}(\mathbb{F}_q)$. Bob chooses an integer d as his private key and publishes $Q = [d]P$ as his public key. Alice selects a random integer k and computes the points $R_1 = [k]P$ and $R_2 = M \oplus [k]Q$ of $\mathcal{E}(\mathbb{F}_q)$. The pair (R_1, R_2) is sent as the ciphertext to Bob. Then Bob can recover the plaintext M by computing $R_2 \oplus [-d]R_1$.

There are some practical problems in implementing an ElGamal elliptic-curve cryptosystem due to the required embedding of the plaintext into $\mathcal{E}(\mathbb{F}_q)$. The reason is that so far there is no convenient method for deterministically generating a large number of points of an elliptic curve. Consequently, several variants of the ElGamal elliptic-curve cryptosystems have been proposed. We describe the variant introduced by Menezes and Vanstone [81].

We start from an elliptic curve $(\mathcal{E}, O)$ over $\mathbb{F}_p$ with p being a prime number and a point $P \in \mathcal{E}(\mathbb{F}_p)$ of large order. The curve $(\mathcal{E}, O)$ and the point P are public knowledge. The message that Alice wants to send to Bob is now assumed to be a pair $(x_1, x_2) \in \mathbb{F}_p^* \times \mathbb{F}_p^*$. Bob chooses an integer d as his private key and publishes $Q = [d]P$ as his public key. Alice selects a random integer k and computes the triple (R, y_1, y_2), where

$$R = [k]P, \quad y_1 = c_1 x_1 \in \mathbb{F}_p^*, \quad y_2 = c_2 x_2 \in \mathbb{F}_p^*,$$

with $(c_1, c_2) = [k]Q$. To decrypt the ciphertext, Bob computes $(y_1 c_1^{-1}, y_2 c_2^{-1})$. Note that Bob can obtain (c_1, c_2) by computing $[d]R$ since $[d]R = [dk]P = [k]Q$. It is obvious that $(y_1 c_1^{-1}, y_2 c_2^{-1}) = (x_1, x_2)$.

It is an important practical aspect of the above two cryptosystems that computing multiples of a point is a major part of the computations involved in encryption and decryption. In recent years, a lot of research toward fast computation of multiples of points of elliptic curves has been carried out (see [9, Chapter IV] and [21]). We present a "toy" example involving multiples of a point of an elliptic curve.

Example 6.2.1. Consider the elliptic curve $(\mathcal{E}, O)$ over $\mathbb{F}_{23}$ defined by $y^2 = x^3 + x + 4$. We have $|\mathcal{E}(\mathbb{F}_{23})| = 29$. It is clear that $P = (0, 2)$ is a point of this curve. Then all points of $\mathcal{E}(\mathbb{F}_{23})$ are given by

$[1]P = (0, 2)$	$[2]P = (13, 12)$	$[3]P = (11, 9)$
$[4]P = (1, 12)$	$[5]P = (7, 20)$	$[6]P = (9, 11)$
$[7]P = (15, 6)$	$[8]P = (14, 5)$	$[9]P = (4, 7)$
$[10]P = (22, 5)$	$[11]P = (10, 5)$	$[12]P = (17, 9)$
$[13]P = (8, 15)$	$[14]P = (18, 9)$	$[15]P = (18, 14)$
$[16]P = (8, 8)$	$[17]P = (17, 14)$	$[18]P = (10, 18)$
$[19]P = (22, 18)$	$[20]P = (4, 16)$	$[21]P = (14, 18)$
$[22]P = (15, 17)$	$[23]P = (9, 12)$	$[24]P = (7, 3)$
$[25]P = (1, 11)$	$[26]P = (11, 14)$	$[27]P = (13, 11)$
$[28]P = (0, 21)$	$[29]P = O$	

The crucial issue concerning the security of the above two cryptosystems is whether the discrete logarithm problem for the group $\mathcal{E}(\mathbb{F}_q)$ of an elliptic curve $(\mathcal{E}, O)$ over $\mathbb{F}_q$ is computationally infeasible. By the Pohlig-Hellman algorithm (see [12, Chapter 9]), if the order of the point $P \in \mathcal{E}(\mathbb{F}_q)$ can be factored into small primes, then the discrete logarithm can be computed easily. Therefore, we usually choose a point P with large prime order. As the order of P divides the order $|\mathcal{E}(\mathbb{F}_q)|$ of $\mathcal{E}(\mathbb{F}_q)$, this forces us to find elliptic curves $(\mathcal{E}, O)$ over $\mathbb{F}_q$ with $|\mathcal{E}(\mathbb{F}_q)|$ divisible by a large prime. If $|\mathcal{E}(\mathbb{F}_q)|$ itself is a large prime number, then the order of any element $\neq O$ of $\mathcal{E}(\mathbb{F}_q)$ is equal to $|\mathcal{E}(\mathbb{F}_q)|$. This requirement raises the question of how to count the number of points of a given elliptic curve. The first polynomial-time algorithm for counting the number of $\mathbb{F}_q$-rational points of a given elliptic curve over $\mathbb{F}_q$ was designed by Schoof [110, 111]. Later this algorithm was improved by Elkies and Atkin (see [30]) and the improved algorithm works well in practice. For details on the elliptic-curve discrete logarithm problem, we refer to [9, Chapter V], [21], and [126, Chapter 5].

In the framework of the discrete logarithm problem for elliptic curves, we can also build key agreement and digital signature schemes (compare with Section 6.1). Let us focus on the following elliptic-curve digital signature scheme.

The Elliptic Curve Digital Signature Algorithm (ECDSA) was adopted in 1999 as an ANSI standard and accepted in 2000 as IEEE and NIST standards. For simplicity, we consider only elliptic curves over $\mathbb{F}_p$ with a large prime p. Then an elliptic curve $(\mathcal{E}, O)$ over $\mathbb{F}_p$ can be defined by a Weierstrass equation

$$y^2 = x^3 + Ax + B$$

for some $A, B \in \mathbb{F}_p$ with $4A^3 + 27B^2 \neq 0$ (see Theorem 3.7.11). Choose a point $P \in \mathcal{E}(\mathbb{F}_p)$ such that the order n of P is a large prime, which is bigger than p. Choose a random integer d and compute the point $Q = [d]P$. All parameters except d are public, while the discrete logarithm d is kept secret. To sign a message $m \in \mathbb{F}_n$, we choose a random $k \in \mathbb{F}_n^*$ and compute $[k]P = (x_1, y_1)$. View $r := x_1$ as an element of $\mathbb{F}_n^*$ and compute $s := k^{-1}(m + dr) \in \mathbb{F}_n$. The pair (r, s) is the signature of the message m. To verify the signature, we do the following:

(i) compute $w = s^{-1} \in \mathbb{F}_n$ (note that if $s = 0$ in $\mathbb{F}_n$, then we choose another random k until we get $s \neq 0$);
(ii) compute $u_1 = mw \pmod{n}$ and $u_2 = rw \pmod{n}$;
(iii) compute $X = [u_1]P \oplus [u_2]Q$;
(iv) if $X = O$, then reject the signature, otherwise convert the x-coordinate x_2 of X to an integer x_2 and compute $v = x_2 \pmod{n}$;
(v) accept the signature if and only if $v = r$.

The above scheme works properly. Indeed, if a signature (r, s) on a message m was generated, then $s = k^{-1}(m + dr) \pmod{n}$. Rearranging gives

$$k \equiv s^{-1}(m + dr) \equiv wm + wrd \equiv u_1 + u_2 d \pmod{n}.$$

Thus, $X = [u_1]P \oplus [u_2]Q = [u_1 + u_2 d]P = [k]P$, and so $v = r$ as required.

We have discussed elliptic-curve cryptography in a relatively brief manner since extensive treatments of this topic are already available in the expository literature. We refer to the handbook edited by Cohen et al. [21] and to the books of Blake, Seroussi, and Smart [9], Enge [32], and Washington [126]. An excellent account of the development of elliptic-curve cryptography up to about 2000 is given in the survey article of Koblitz, Menezes, and Vanstone [64].

6.3 Hyperelliptic-Curve Cryptography

As we pointed out in Section 6.1, various cryptographic schemes can be designed by using suitable groups containing large finite cyclic subgroups.

An interesting candidate is the group $\mathrm{Cl}(F) = \mathrm{Div}^0(F)/\mathrm{Princ}(F)$ of divisor classes of degree 0 of a global function field F. We recall that $\mathrm{Cl}(F)$ is a finite abelian group and that its order $|\mathrm{Cl}(F)|$ is the divisor class number of F (see Section 4.1). According to Theorem 4.2.5, we have

$$(q^{1/2} - 1)^{2g} \leq |\mathrm{Cl}(F)| \leq (q^{1/2} + 1)^{2g},$$

where q is the order of the full constant field of F and g is the genus of F.

To be of practical interest for cryptography, the global function field F has to be chosen in such a way that $|\mathrm{Cl}(F)|$ can be computed with a reasonable effort and that there exists an efficient and conveniently implementable algorithm for carrying out the group operation in $\mathrm{Cl}(F)$. This imposes a considerable restriction on the choice of global function fields, or equivalently of algebraic curves over finite fields.

The case of elliptic curves over finite fields was already discussed in Section 6.2. A fairly wide class of potentially useful curves is that of C_{ab} curves. Let a and b be two positive coprime integers. A C_{ab} curve is a curve over $\mathbb{F}_q$ with an affine equation of the form

$$\sum_{\substack{i \geq 0,\ j \geq 0 \\ ai + bj \leq ab}} \alpha_{i,j} x^i y^j = 0,$$

where all $\alpha_{i,j} \in \mathbb{F}_q$ and $\alpha_{b,0}$ and $\alpha_{0,a}$ are nonzero. Several authors have investigated algorithmic aspects for this general class of curves. For instance, Arita [1] designed an algorithm for adding elements in the group $\mathrm{Cl}(F)$ when F is the function field of a C_{ab} curve.

An interesting special family of C_{ab} curves is that of Picard curves. A Picard curve is a nonsingular projective curve over $\mathbb{F}_q$ of genus 3 with an affine equation of the form $y^3 = f(x)$, where $f \in \mathbb{F}_q[x]$ is a polynomial of degree 4. A detailed discussion of relevant algorithms for Picard curves can be found in Bauer, Teske, and Weng [5].

Among C_{ab} curves of genus larger than 1, the greatest attention has been devoted to hyperelliptic curves. The use of hyperelliptic curves, or equivalently of hyperelliptic function fields, for cryptography was first proposed by Koblitz [63]. For simplicity, we discuss only the case where q is odd. For a given integer $g \geq 2$, we consider the defining equation of a *hyperelliptic function field* $F - \mathbb{F}_q(x, y)$ of genus g in the normalized form $y^2 - f(x)$, where $f \in \mathbb{F}_q[x]$ is a monic and squarefree polynomial of degree $2g+1$. This form is always assumed when we speak of a hyperelliptic function field.

As mentioned before, for practical cryptographic purposes we need to be able to carry out the addition in $\mathrm{Cl}(F)$ in an efficient manner. This can indeed be achieved in the case of a hyperelliptic function field F. The first step is a convenient representation of the elements of $\mathrm{Cl}(F)$. Mumford [84] gave the following representation, which we state without proof.

Proposition 6.3.1. Let $F = \mathbb{F}_q(x, y)$ be a hyperelliptic function field of genus g with defining equation $y^2 = f(x)$ as above. Then there is a one-to-one correspondence between the elements of $\mathrm{Cl}(F)$ and the pairs (u, v) of elements of $\mathbb{F}_q[x]$ with u monic, u dividing $f - v^2$, and $\deg(v) < \deg(u) \le g$.

Remark 6.3.2. In this representation, the zero element of $\mathrm{Cl}(F)$ corresponds to the pair $(1, 0)$. If an element of $\mathrm{Cl}(F)$ is represented by (u, v), then the inverse element in this group is represented by $(u, -v)$.

The addition in $\mathrm{Cl}(F)$ can be implemented by describing how two pairs (u_1, v_1) and (u_2, v_2) in Proposition 6.3.1 are added. An efficient algorithm for this operation was first designed by Cantor [16]. This algorithm starts out by computing $d = \gcd(u_1, u_2, v_1 + v_2)$. The greatest common divisor d can be represented as a linear combination of u_1, u_2, and $v_1 + v_2$. Thus, there exist polynomials $a_1, a_2, a_3 \in \mathbb{F}_q[x]$ such that

$$d = a_1 u_1 + a_2 u_2 + a_3(v_1 + v_2).$$

The relevant computations can be carried out by the (extended) Euclidean algorithm for polynomials over $\mathbb{F}_q$. Now we put

$$u_0 = \frac{u_1 u_2}{d^2}, \qquad v_0 \equiv \frac{a_1 u_1 v_2 + a_2 u_2 v_1 + a_3(v_1 v_2 + f)}{d} \pmod{u_0},$$

where $\deg(v_0) < \deg(u_0)$. If $\deg(u_0) \le g$, then the pair (u_0, v_0) represents the sum of (u_1, v_1) and (u_2, v_2). Otherwise we have $g < \deg(u_0) \le 2g$, and then some reduction steps are needed, which are similar to the reduction of binary quadratic forms. We set $b_0 = u_0$, $c_0 = v_0$, and for $i = 1, 2, \ldots$ we put

$$b_i = \frac{f - c_{i-1}^2}{b_{i-1}}, \qquad c_i \equiv -c_{i-1} \pmod{b_i},$$

where $\deg(c_i) < \deg(b_i)$. This procedure is stopped as soon as we reach an index k with $\deg(b_k) \le g$. Let $\overline{b_k}$ be the monic polynomial obtained by dividing b_k by its leading coefficient. Then the pair $(u, v) := (\overline{b_k}, c_k)$ represents the sum of (u_1, v_1) and (u_2, v_2).

Example 6.3.3. Let $q = 3$, $g = 2$, $f(x) = x^5 - x^4 - x \in \mathbb{F}_3[x]$, and let $F = \mathbb{F}_3(x, y)$ be the hyperelliptic function field defined by $y^2 = f(x)$. The pairs $(u_1, v_1) = (x^2 - x - 1, x)$ and $(u_2, v_2) = (x, 0)$ are of the form in Proposition 6.3.1 and thus represent elements of $\mathrm{Cl}(F)$. To compute the sum of (u_1, v_1) and (u_2, v_2), we first note that $d = \gcd(u_1, u_2, v_1 + v_2) = 1$. Furthermore,

$$1 = (-1) \cdot (x^2 - x - 1) + (x - 1) \cdot x,$$

so that we can take $a_1(x) = -1$, $a_2(x) = x - 1$, $a_3(x) = 0$. This yields $u_0(x) = x^3 - x^2 - x$ and $v_0(x) \equiv x^3 - x^2 \pmod{x^3 - x^2 - x}$, hence $v_0(x) = x$. Since $\deg(u_0) > g$, we need to go through the reduction procedure. We have $b_0(x) = x^3 - x^2 - x$ and $c_0(x) = x$, and then

$$b_1(x) = \frac{x^5 - x^4 - x^2 - x}{x^3 - x^2 - x} = x^2 + 1, \qquad c_1(x) = -x.$$

The pair $(u, v) = (x^2 + 1, -x)$ satisfies the conditions in Proposition 6.3.1 and therefore represents the sum of (u_1, v_1) and (u_2, v_2) in $\mathrm{Cl}(F)$.

Cantor [16] already investigated how to speed up the reduction steps in the above algorithm for addition in $\mathrm{Cl}(F)$. A detailed analysis of improved reduction methods was presented by Stein [116]. In the important special case of hyperelliptic function fields F of genus 2, explicit formulas for performing the arithmetic in $\mathrm{Cl}(F)$, which lead to further speedups, were provided by Lange [67].

Besides fast arithmetic in the group $\mathrm{Cl}(F)$, another major issue is the computation of the order of $\mathrm{Cl}(F)$, that is, of the divisor class number $h(F)$ of the hyperelliptic function field F. To simplify this task, it has been suggested in the literature to start from a hyperelliptic function field over a relatively small finite field and then choose F to be a constant field extension of large degree. Then Theorem 4.1.11 and the methods in Section 4.1 can be used to facilitate the computation of $h(F)$.

From the viewpoint of the security of the corresponding cryptographic schemes, the hyperelliptic function field F has to be chosen in such a way that the discrete logarithm problem for $\mathrm{Cl}(F)$ is computationally infeasible.

Some of the remarks we made on the security of elliptic-curve cryptosystems in Section 6.2 apply also to the present case. There are several powerful algorithms for computing discrete logarithms in $\mathrm{Cl}(F)$ that are specifically tailored to hyperelliptic function fields, such as the index-calculus algorithms of Enge [33] and Gaudry [39]. These algorithms indicate that within the family of hyperelliptic function fields, only those of small genus should be used for cryptographic purposes. The remarkable work initiated by Frey [34] and continued by Gaudry, Hess, and Smart [40] shows how to use Weil descent to reduce certain instances of the discrete logarithm problem for elliptic curves to a discrete logarithm problem for hyperelliptic curves of large genus.

Detailed discussions of the security of cryptographic schemes based on hyperelliptic curves (or equivalently hyperelliptic function fields) and of many other issues related to these curves and function fields can be found in the handbook edited by Cohen et al. [21].

6.4 Code-Based Public-Key Cryptosystems

There are interesting links between coding theory and cryptography, which mostly stem from the observation that coding theory is a source of hard problems that can form the basis of cryptosystems. We use the standard terminology and notation from coding theory introduced in Section 5.1.

A public-key cryptosystem using linear codes is the *McEliece cryptosystem* proposed in [79]. Let C be a linear $[n, k, d]$ code over $\mathbb{F}_q$ and let the $k \times n$ matrix G be a *generator matrix* of C, that is, G is such that C is the row space of G. The matrix G is part of the private key of Bob. Next, Bob chooses two more matrices. The matrix S is an arbitrary nonsingular $k \times k$ matrix over $\mathbb{F}_q$. Furthermore, P is an $n \times n$ matrix over $\mathbb{F}_q$, which is obtained from a nonsingular diagonal matrix by arbitrary row permutations. The matrices G, S, and P form Bob's private key. The public key of Bob is the $k \times n$ matrix $G' := SGP$ which may be viewed as a scrambled version of G.

The admissible plaintexts in the McEliece cryptosystem are row vectors $\mathbf{x} \in \mathbb{F}_q^k$. If Alice wants to encrypt the plaintext $\mathbf{x} \in \mathbb{F}_q^k$ destined for Bob, she chooses a random row vector $\mathbf{z} \in \mathbb{F}_q^n$ with weight $w(\mathbf{z}) \leq t := \lfloor (d-1)/2 \rfloor$ and uses Bob's public key G' to compute the ciphertext $\mathbf{y} = \mathbf{x}G' + \mathbf{z} \in \mathbb{F}_q^n$. Note that the condition $w(\mathbf{z}) \leq t$ means that $\mathbf{z}$ can be regarded as an error vector that the code C is able to correct. This observation is used in the decryption algorithm. Concretely, if Bob receives the ciphertext $\mathbf{y}$, then he first computes

$\mathbf{y}' = \mathbf{y}P^{-1} = \mathbf{x}SG + \mathbf{z}P^{-1}$. Now

$$w(\mathbf{y}' - \mathbf{x}SG) = w(\mathbf{z}P^{-1}) = w(\mathbf{z}) \leq t.$$

Since $\mathbf{x}SG = (\mathbf{x}S)G$ belongs to the code C, the vector $\mathbf{y}'$ is like a received word that can be corrected by C to produce the original word $\mathbf{x}S$. From this, Bob recovers the plaintext $\mathbf{x} = (\mathbf{x}S)S^{-1}$.

A related public-key cryptosystem is the *Niederreiter cryptosystem* introduced in [86]. This is a cryptosystem for low-weight plaintext vectors. The starting point in the design of this cryptosystem is again a linear $[n, k, d]$ code C over $\mathbb{F}_q$. As part of his private key, Bob chooses a parity-check matrix H of C, which is an $(n - k) \times n$ matrix over $\mathbb{F}_q$. Furthermore, Bob selects a matrix P as in the McEliece cryptosystem as well as an arbitrary nonsingular $(n - k) \times (n - k)$ matrix M over $\mathbb{F}_q$. The matrices H, M, and P form the private key of Bob. His public key is the $(n - k) \times n$ matrix $H' := MHP$, which may be regarded as a scrambled version of H.

The admissible plaintexts in the Niederreiter cryptosystem are column vectors $\mathbf{x} \in \mathbb{F}_q^n$ with weight $w(\mathbf{x}) \leq t := \lfloor (d - 1)/2 \rfloor$. Alice encrypts $\mathbf{x} \in \mathbb{F}_q^n$ by computing the ciphertext $\mathbf{y} = H'\mathbf{x}$ using Bob's public key H'. Upon receiving $\mathbf{y}$, which is a column vector from $\mathbb{F}_q^{n-k}$, Bob first computes $\mathbf{y}' = M^{-1}\mathbf{y} = HP\mathbf{x}$. Now $\mathbf{x}' := P\mathbf{x}$ satisfies $w(\mathbf{x}') \leq t$, and so $\mathbf{x}'$ can be viewed as an error vector that the code C is able to correct. Thus, Bob can recover $\mathbf{x}'$ uniquely from $\mathbf{y}' = H\mathbf{x}'$. Finally, Bob obtains the plaintext $\mathbf{x} = P^{-1}\mathbf{x}'$.

The following result of Li, Deng, and Wang [70] shows that, with corresponding choices of code parameters, the McEliece and Niederreiter cryptosystems have equivalent security levels, in the sense that the problem of breaking either cryptosystem can be reduced to that of breaking the other cryptosystem by a polynomial-time algorithm.

Theorem 6.4.1. With corresponding choices of code parameters, the McEliece cryptosystem and the Niederreiter cryptosystem have equivalent security levels.

Proof. Suppose first that we have an algorithm to break the McEliece cryptosystem. This means that given a $k \times n$ generator matrix G' of a linear code C' and a row vector $\mathbf{y} = \mathbf{x}G' + \mathbf{z}$ with $w(\mathbf{z}) < t$, the algorithm determines the row vector $\mathbf{x}$. Now consider an instance of the Niederreiter cryptosystem, that is, we are given an $(n - k) \times n$

parity-check matrix H' of C' and a column vector $\mathbf{y}' = H'\mathbf{x}'$ with $w(\mathbf{x}') \leq t$. Note that by using linear algebra we can compute a basis of the solution space of the system of linear equations $H'\mathbf{v} = \mathbf{0}$ and so a generator matrix G' of C', and this can be done in polynomial time. Next we find any column vector $\mathbf{u}$, which solves $H'\mathbf{u} = \mathbf{y}'$, and this is again a polynomial-time algorithm. Then we have $H'(\mathbf{u}-\mathbf{x}') = \mathbf{0}$, hence $\mathbf{u} - \mathbf{x}' \in C'$. Therefore, $(\mathbf{u} - \mathbf{x}')^{\mathrm{T}} = \mathbf{a}G'$ for some row vector $\mathbf{a} \in \mathbb{F}_q^k$, or equivalently $\mathbf{u}^{\mathrm{T}} = \mathbf{a}G' + (\mathbf{x}')^{\mathrm{T}}$ with $w(\mathbf{x}') \leq t$. Thus, the algorithm for breaking the McEliece cryptosystem can be applied and yields $\mathbf{a}$, and so $\mathbf{x}'$ from $(\mathbf{x}')^{\mathrm{T}} = \mathbf{u}^{\mathrm{T}} - \mathbf{a}G'$. Hence, we have broken the given instance of the Niederreiter cryptosystem.

Conversely, suppose that we know an algorithm to break the Niederreiter cryptosystem. This means that given an $(n - k) \times n$ parity-check matrix H' of a linear code C' and a column vector $\mathbf{y}' = H'\mathbf{x}'$ with $w(\mathbf{x}') \leq t$, the algorithm determines the column vector $\mathbf{x}'$. Now consider an instance of the McEliece cryptosystem, that is, we are given a $k \times n$ generator matrix G' of C' and a row vector $\mathbf{y}$, which is a ciphertext. Again by linear algebra, we can compute a parity-check matrix H' of C' in polynomial time. Now $\mathbf{y} = \mathbf{x}G' + \mathbf{z}$ with $w(\mathbf{z}) \leq t$, and so

$$\mathbf{y}(H')^{\mathrm{T}} = \mathbf{x}G'(H')^{\mathrm{T}} + \mathbf{z}(H')^{\mathrm{T}} = \mathbf{z}(H')^{\mathrm{T}}.$$

Therefore, $H'\mathbf{z}^{\mathrm{T}} = H'\mathbf{y}^{\mathrm{T}}$ with $w(\mathbf{z}^{\mathrm{T}}) \leq t$. Thus, the algorithm for breaking the Niederreiter cryptosystem can be applied and returns the vector $\mathbf{z}^{\mathrm{T}}$. Then we can compute $\mathbf{x}G' = \mathbf{y} - \mathbf{z}$, and this yields the row vector $\mathbf{x}$ in polynomial time. Hence we have broken the given instance of the McEliece cryptosystem. □

In view of Theorem 6.4.1, the security of the McEliece and Niederreiter cryptosystems can be analyzed simultaneously. Canteaut and Sendrier [15] conclude, after a detailed cryptanalysis of both cryptosystems, that currently secure parameters for the binary case $q = 2$ are given by $n = 2048, k = 1608$, and $t = 40$. They also point out that, with these parameters, the encryption and decryption for the Niederreiter cryptosystem are much faster than for the RSA cryptosystem (see Example 6.1.1) with comparable parameters. The paper of Johansson and Jönsson [61] and the references therein contain further discussions of attacks on code-based cryptosystems.

Berson [6] noticed a weakness of the McEliece cryptosystem, which is not present in the Niederreiter cryptosystem, namely that the McEliece cryptosystem can be subjected to the message-resend attack. Suppose that the attacker obtains two different ciphertexts $\mathbf{y}_1$ and $\mathbf{y}_2$, which both correspond to the same plaintext $\mathbf{x}$. That is, we have $\mathbf{y}_1 = \mathbf{x}G' + \mathbf{z}_1$ and $\mathbf{y}_2 = \mathbf{x}G' + \mathbf{z}_2$ with $w(\mathbf{z}_1) \leq t$ and $w(\mathbf{z}_2) \leq t$. Let

$$\mathbf{b} = (b_1, \ldots, b_n) = \mathbf{y}_1 - \mathbf{y}_2 = \mathbf{z}_1 - \mathbf{z}_2.$$

Put $B = \{1 \leq j \leq n : b_j = 0\}$. The attack is based on decoding algorithms and on the fact that most elements of B are error-free locations of both $\mathbf{y}_1$ and $\mathbf{y}_2$. This attack cannot be mounted against the Niederreiter cryptosystem since the encryption in this cryptosystem is deterministic.

The Niederreiter cryptosystem shows a reasonable behavior with respect to the information rate, although its set of plaintexts is restricted by the weight condition. Let $S(\mathcal{C})$ be the number of possible plaintexts and $T(\mathcal{C})$ the number of possible ciphertexts in a cryptosystem $\mathcal{C}$. Then the *information rate* of the cryptosystem $\mathcal{C}$ is defined by

$$R(\mathcal{C}) = \frac{\log S(\mathcal{C})}{\log T(\mathcal{C})}.$$

The information rate $R(\mathcal{C})$ may be viewed as the amount of information contained per bit of ciphertext. In the following theorem, we use the q-ary entropy function H_q defined in Section 5.3.

Theorem 6.4.2. Let q be an arbitrary prime power and let $0 < \theta < (q - 1)/2q$. Then there exists a sequence $\mathcal{C}_1, \mathcal{C}_2, \ldots$ of Niederreiter cryptosystems based on linear codes over $\mathbb{F}_q$ with lengths $n(\mathcal{C}_i) \to \infty$ as $i \to \infty$ and weight conditions $w(\mathbf{x}) \leq t(\mathcal{C}_i)$ on the plaintexts $\mathbf{x}$ such that

$$\lim_{i\to\infty} \frac{t(\mathcal{C}_i)}{n(\mathcal{C}_i)} = \theta, \qquad \lim_{i\to\infty} R(\mathcal{C}_i) \geq \frac{H_q(\theta)}{H_q(2\theta)}.$$

Proof. By the asymptotic Gilbert-Varshamov bound in Theorem 5.3.2, there exists a sequence $\mathcal{C}_1, \mathcal{C}_2, \ldots$ such that each C_i is a linear $[n_i, k_i, d_i]$ code over $\mathbb{F}_q$ and we have $n_i \to \infty$ as $i \to \infty$ as well as

$$\lim_{i\to\infty} \frac{d_i}{n_i} = 2\theta, \qquad \lim_{i\to\infty} \frac{k_i}{n_i} \geq 1 - H_q(2\theta).$$

For each $i = 1, 2, \ldots$, let C_i be a Niederreiter cryptosystem based on the linear code C_i. Then for $i = 1, 2, \ldots$, we have $n(C_i) = n_i$ and

$$t(C_i) = \left\lfloor \frac{d_i - 1}{2} \right\rfloor.$$

Therefore,

$$\lim_{i \to \infty} \frac{t(C_i)}{n(C_i)} = \theta.$$

Furthermore, we have

$$S(C_i) = \sum_{j=0}^{t(C_i)} \binom{n_i}{j} (q-1)^j \quad \text{for } i = 1, 2, \ldots,$$

and then by adapting an argument in the proof of Theorem 5.3.2 we can show that

$$\lim_{i \to \infty} \frac{1}{n_i} \log_q S(C_i) = H_q(\theta).$$

Moreover, $T(C_i) = q^{n_i - k_i}$ for $i = 1, 2, \ldots$, and so

$$\lim_{i \to \infty} \frac{1}{n_i} \log_q T(C_i) = 1 - \lim_{i \to \infty} \frac{k_i}{n_i} \leq H_q(2\theta).$$

It follows that

$$\lim_{i \to \infty} R(C_i) = \lim_{i \to \infty} \frac{\frac{1}{n_i} \log_q S(C_i)}{\frac{1}{n_i} \log_q T(C_i)} \geq \frac{H_q(\theta)}{H_q(2\theta)},$$

which completes the proof of the theorem.   □

Example 6.4.3. Consider the case $q = 2$ in Theorem 6.4.2. Since $H_2(\frac{1}{4}) > 0.81$ and $H_2(\frac{1}{2}) = 1$, there exists a choice of θ sufficiently close to $\frac{1}{4}$ such that we obtain

$$\lim_{i \to \infty} R(C_i) \geq 0.81.$$

Thus, asymptotically more than 80% of the ciphertext in the Nieder-reiter cryptosystems $\mathcal{C}_1, \mathcal{C}_2, \ldots$ in the proof of Theorem 6.4.2 carries information.

We remark that, as a consequence of Theorem 5.3.7, the lower bound on the asymptotic information rate of the cryptosystems $\mathcal{C}_1, \mathcal{C}_2, \ldots$ in Theorem 6.4.2 can be achieved by using Niederreiter cryptosystems based on a sequence of algebraic-geometry codes. Furthermore, the improvements on the asymptotic Gilbert-Varshamov bound for linear codes discussed in Section 5.3 yield corresponding improved lower bounds on the asymptotic information rates of Niederreiter cryptosystems.

Unlike the RSA cryptosystem, the McEliece and Niederreiter cryptosystems cannot readily be converted into digital signature schemes. However, a digital signature scheme that is based on an adaptation of the Niederreiter cryptosystem was designed by Courtois, Finiasz, and Sendrier [22] and turns out to be quite practical in terms of length of the signature and cost of the verification procedure. Further connections between coding theory and cryptography are discussed in [89].

6.5 Frameproof Codes

Intellectual property protection is a critical issue in today's world. In order to protect copyrighted material (computer software, videos, etc.), vendors may endow each copy of the material with some "fingerprint" that allows them to detect any unauthorized copy and trace this copy back to the user who created it.

Let us assume that the fingerprint is a vector $(a_1, \ldots, a_n) \in \mathbb{F}_q^n$ of length n. If an unauthorized copy of the material is found containing the fingerprint of user U, then we can accuse U of having produced a pirated copy. However, U could claim that he/she was framed by a coalition of other users who created the fingerprint of U from components of their own fingerprints. Thus, it is desirable to construct fingerprint schemes that satisfy the following property: no coalition of users can collude to frame a user not in the coalition. Schemes satisfying this property are called frameproof codes, and if the condition is relaxed by limiting the size of a coalition to at most c users, then we speak of a c-frameproof code. We emphasize that the codes in this section are not used for error correction, but for fingerprinting.

Before introducing the formal definition of a c-frameproof code, we define for each $i = 1, \ldots, n$ the projection $\pi_i : \mathbb{F}_q^n \to \mathbb{F}_q$ by

$$\pi_i(\mathbf{x}) = x_i \quad \text{for } \mathbf{x} = (x_1, \ldots, x_n) \in \mathbb{F}_q^n.$$

Furthermore, for a subset A of $\mathbb{F}_q^n$, we define the set $\mathrm{desc}(A)$ of descendants of A to be the set of all $\mathbf{x} \in \mathbb{F}_q^n$ such that for each $i = 1, \ldots, n$ there exists a vector $\mathbf{a}_i \in A$ satisfying $\pi_i(\mathbf{x} - \mathbf{a}_i) = 0$. Note that $\mathrm{desc}(A)$ is the set of all vectors in $\mathbb{F}_q^n$ that can be formed from corresponding components of vectors in A.

Definition 6.5.1. Let q be a prime power and let $c \geq 2$ and $n \geq 2$ be integers. Then a *c-frameproof code* over $\mathbb{F}_q$ of length n is a subset C of $\mathbb{F}_q^n$ such that for all $A \subseteq C$ with $|A| \leq c$ we have

$$\mathrm{desc}(A) \cap C = A.$$

It is clear from this definition that a c-frameproof code over $\mathbb{F}_q$ of length n is also a c_1-frameproof code over $\mathbb{F}_q$ of length n for any integer c_1 with $2 \leq c_1 \leq c$.

Example 6.5.2. The following is a simple construction of a c-frameproof code. For $n \geq 2$ and any prime power q, let C be the set of all vectors in $\mathbb{F}_q^n$ of weight exactly 1. Then $|C| = (q - 1)n$. Now let A be an arbitrary nonempty subset of C and take $\mathbf{b} \in \mathrm{desc}(A) \cap C$. Suppose the ith component b_i of $\mathbf{b}$ is nonzero. Since $\mathbf{b} \in \mathrm{desc}(A)$, it follows that there exists $\mathbf{a} = (a_1, \ldots, a_n) \in A$ with $a_i = b_i$. But all other components of $\mathbf{b}$ and $\mathbf{a}$ are 0, hence $\mathbf{b} = \mathbf{a} \in A$. Thus, C is a c-frameproof code over $\mathbb{F}_q$ of length n for any integer $c \geq 2$.

For practical purposes, we would like to make the size of a c-frameproof code over $\mathbb{F}_q$ of length n as large as possible for given q, c, and n. This leads to the following definition of the quantity $M_q(c, n)$. In the sequel, we prove an upper bound on $M_q(c, n)$ due to Blackburn [8].

Definition 6.5.3. For a given prime power q and integers $c \geq 2$ and $n \geq 2$, let $M_q(c, n)$ denote the largest cardinality of a c-frameproof code over $\mathbb{F}_q$ of length n.

Theorem 6.5.4. Let q be a prime power and let $c \geq 2$ and $n \geq 2$ be integers. Then

$$M_q(c, n) \leq \max \left\{ q^{\lceil n/c \rceil}, r \left(q^{\lceil n/c \rceil} - 1 \right) + (c - r) \left(q^{\lfloor n/c \rfloor} - 1 \right) \right\},$$

where r is the least residue of n modulo c.

Proof. Let C be a c-frameproof code over $\mathbb{F}_q$ of length n. For any subset S of $\{1, \ldots, n\}$, we define

$$V_S = \{ \mathbf{x} \in C : \text{there is no } \mathbf{y} \in C \setminus \{\mathbf{x}\} \text{ such that } x_i = y_i \quad \text{for all } i \in S \},$$

where we write $\mathbf{x} = (x_1, \ldots, x_n)$ and $\mathbf{y} = (y_1, \ldots, y_n)$. Note that by the definition of V_S, every $\mathbf{x} \in V_S$ is uniquely determined by $(x_i : i \in S)$, and so $|V_S| \leq q^{|S|}$. Moreover, if $|C| > q^{|S|}$, then $|V_S| \leq q^{|S|} - 1$ since at least one choice of $(x_i : i \in S)$ must correspond to two or more elements of C.

Now we choose c disjoint subsets $S_1, \ldots, S_c$ of $\{1, \ldots, n\}$ in such a way that $|S_j| = \lceil n/c \rceil$ for $1 \leq j \leq r$ and $|S_j| = \lfloor n/c \rfloor$ for $r + 1 \leq j \leq c$. Then

$$\bigcup_{j=1}^{c} S_j = \{1, \ldots, n\}.$$

The bound of the theorem follows if we can show that

$$C = \bigcup_{j=1}^{c} V_{S_j}.$$

Suppose, on the contrary, that there exists a vector $\mathbf{z} \in C \setminus \bigcup_{j=1}^{c} V_{S_j}$. Then, for each $j = 1, \ldots, c$, there is a $\mathbf{y}_j \in C \setminus \{\mathbf{z}\}$ such that $\mathbf{y}_j$ and $\mathbf{z}$ agree in their ith components for all $i \in S_j$. But then $\mathbf{z} \in \text{desc}(\{\mathbf{y}_1, \ldots, \mathbf{y}_c\})$, which contradicts the fact that C is a c-frameproof code. This contradiction completes the proof. $\square$

Corollary 6.5.5. For $2 \leq n \leq c$ we have $M_q(c, n) = (q - 1)n$ for any prime power q.

Proof. From Theorem 6.5.4 we obtain $M_q(c, n) \leq (q-1)n$ if $2 \leq n \leq c$. On the other hand, Example 6.5.2 shows that $M_q(c, n) \geq (q - 1)n$. □

Theorem 6.5.4 suggests to consider the asymptotic quantity $D_q(c)$ defined as follows. For a given prime power q and an integer $c \geq 2$, let

$$D_q(c) := \limsup_{n \to \infty} \frac{\log_q M_q(c, n)}{n}.$$

The following result is an immediate consequence of Theorem 6.5.4.

Corollary 6.5.6. We always have

$$D_q(c) \leq \frac{1}{c}.$$

It is an important task to find lower bounds on $D_q(c)$, which come as close as possible to $\frac{1}{c}$. For this purpose, we need to construct long c-frameproof codes over $\mathbb{F}_q$ of large size. The following general construction principle based on error-correcting codes was introduced by Cohen and Encheva [20].

Proposition 6.5.7. Let q be a prime power and let $n \geq 3$ be an integer. Then any code over $\mathbb{F}_q$ of length n and minimum distance d, where $(n + 1)/2 \leq d < n$, is a c-frameproof code over $\mathbb{F}_q$ of length n with $c = \lfloor (n - 1)/(n - d) \rfloor$.

Proof. Let C be a given code over $\mathbb{F}_q$ of length n and minimum distance d, where $(n + 1)/2 \leq d < n$. Let A be a subset of C with $|A| \leq c = \lfloor (n - 1)/(n - d) \rfloor$. Suppose A were a proper subset of $\mathrm{desc}(A) \cap C$ and choose $\mathbf{x} \in (\mathrm{desc}(A) \cap C) \setminus A$. Since $|A| \leq c$ and $\mathbf{x} \in \mathrm{desc}(A)$, it follows that there is a $\mathbf{y} \in A$, which agrees with $\mathbf{x}$ in at least $\lceil n/c \rceil$ components. Thus, the weight of $\mathbf{y} - \mathbf{x}$ satisfies

$$w(\mathbf{y} - \mathbf{x}) \leq n - \left\lceil \frac{n}{c} \right\rceil \leq n - \frac{n}{c}.$$

But $\mathbf{x}, \mathbf{y} \in C$ and $\mathbf{x} \neq \mathbf{y}$, hence,

$$d \leq w(\mathbf{y} - \mathbf{x}) \leq n - \frac{n}{c},$$

which implies $c \geq n/(n - d)$. This contradicts $c = \lfloor (n - 1)/(n - d) \rfloor$. □

The above relationship between error-correcting codes and frameproof codes, when combined with the asymptotic Gilbert-Varshamov bound in Theorem 5.3.2, leads to the following lower bound on $D_q(c)$.

Theorem 6.5.8. If q is a prime power and c an integer with $2 \le c \le q$, then

$$D_q(c) \ge 1 - H_q \left(1 - \frac{1}{c} \right),$$

where H_q is the q-ary entropy function.

Proof. Since $H_q(1 - \frac{1}{q}) = 1$, the result is trivial for $c = q$, and so we can assume that $2 \le c \le q - 1$. Choose $\varepsilon > 0$ so small that

$$\delta := 1 - \frac{1}{c} + \varepsilon < 1 - \frac{1}{c+1}.$$

By Theorem 5.3.2, there exists a sequence $C_1, C_2, \ldots$ of codes over $\mathbb{F}_q$ with lengths $n(C_i) \to \infty$ as $i \to \infty$ and such that

$$\lim_{i \to \infty} \frac{d(C_i)}{n(C_i)} = \delta, \qquad \lim_{i \to \infty} \frac{\log_q |C_i|}{n(C_i)} \ge 1 - H_q(\delta).$$

The definitions of ε and δ imply that for sufficiently large i we have

$$1 - \frac{n(C_i) - 1}{c n(C_i)} < \frac{d(C_i)}{n(C_i)} < 1 - \frac{n(C_i) - 1}{(c+1)n(C_i)}.$$

These inequalities are equivalent to

$$c \le \frac{n(C_i) - 1}{n(C_i) - d(C_i)} < c + 1.$$

It follows then from Proposition 6.5.7 that C_i is a c-frameproof code over $\mathbb{F}_q$ for sufficiently large i. Consequently,

$$D_q(c) \ge \lim_{i \to \infty} \frac{\log_q |C_i|}{n(C_i)} \ge 1 - H_q \left(1 - \frac{1}{c} + \varepsilon \right).$$

Letting ε tend to 0, we obtain the desired result. $\square$

Further lower bounds on $D_q(c)$ can be derived from the asymptotic theory of codes. For instance, the following result is a consequence of the TVZ bound in Theorem 5.3.8. The quantity $A(q)$ is given by Definition 4.3.6.

Theorem 6.5.9. For any prime power q and any integer $c \geq 2$, we have

$$D_q(c) \geq \frac{1}{c} - \frac{1}{A(q)}.$$

Proof. This is shown in a similar way as Theorem 6.5.8, but using Theorem 5.3.8 instead of Theorem 5.3.2. ☐

By combining Corollary 6.5.6 and Theorem 6.5.9, we obtain

$$\frac{1}{c} - \frac{1}{A(q)} \leq D_q(c) \leq \frac{1}{c}.$$

A result mentioned in Remark 4.3.8 shows that $A(q)$ is at least of the order of magnitude $\log q$, and so $\frac{1}{A(q)} \to 0$ as $q \to \infty$. Consequently, $D_q(c)$ converges to $\frac{1}{c}$ as $q \to \infty$ for any fixed integer $c \geq 2$.

We now describe a direct coding-theoretic approach to a lower bound on $D_q(c)$ which does not rely on Proposition 6.5.7. We use a variant of the construction of algebraic-geometry codes in Section 5.2 to obtain a frameproof code over $\mathbb{F}_q$ of given length n. Let $F/\mathbb{F}_q$ be a global function field with $N(F) \geq n$ and let $P_1, \ldots, P_n$ be n distinct rational places of F. Choose a positive divisor G of F such that $\mathcal{L}(G - \sum_{i=1}^{n} P_i) = \{0\}$. For each $i = 1, \ldots, n$, put $v_i := v_{P_i}(G) \geq 0$ and let $t_i \in F$ be a local parameter at P_i. Now we introduce the $\mathbb{F}_q$-linear map $\psi : \mathcal{L}(G) \to \mathbb{F}_q^n$ given by

$$\psi(f) = \left((t_1^{v_1} f)(P_1), \ldots, (t_n^{v_n} f)(P_n) \right) \quad \text{for all } f \in \mathcal{L}(G).$$

The image $C(\sum_{i=1}^{n} P_i, G)$ of ψ is an $\mathbb{F}_q$-linear subspace of $\mathbb{F}_q^n$. Note that the condition $\mathcal{L}(G - \sum_{i=1}^{n} P_i) = \{0\}$ implies that ψ is injective, and so the dimension of $C(\sum_{i=1}^{n} P_i, G)$ is equal to $\mathcal{L}(G)$.

Proposition 6.5.10. Let $F/\mathbb{F}_q$ be a global function field with $N(F) \geq n \geq 2$ and let $P_1, \ldots, P_n$ be n distinct rational places of F. Let G be a positive divisor of F with $\deg(G) < n$. Let the integer $c \geq 2$ satisfy $\mathcal{L}(cG - \sum_{i=1}^{n} P_i) = \{0\}$. Then $C(\sum_{i=1}^{n} P_i, G)$ is a c-frameproof code over $\mathbb{F}_q$ of length n with cardinality $q^{\ell(G)}$.

Proof. It suffices to show the property of a c-frameproof code in Definition 6.5.1. Let $A = \{\psi(f_1), \ldots, \psi(f_r)\}$ be a subset of $C(\sum_{i=1}^{n} P_i, G)$ with $r := |A|$ satisfying $1 \leq r \leq c$. Let $\psi(h) \in \operatorname{desc}(A) \cap C(\sum_{i=1}^{n} P_i, G)$ for some $h \in \mathcal{L}(G)$. Then by the definition of descendant, for each $i = 1, \ldots, n$ we have

$$\prod_{j=1}^{r} \pi_i(\psi(f_j) - \psi(h)) = 0.$$

This implies that

$$\prod_{j=1}^{r}(t_i^{v_i} f_j - t_i^{v_i} h)(P_i) = 0,$$

that is,

$$v_{P_i}\left(\prod_{j=1}^{r}(t_i^{v_i} f_j - t_i^{v_i} h)\right) \geq 1.$$

This is equivalent to

$$v_{P_i}\left(\prod_{j=1}^{r}(f_j - h)\right) \geq -rv_i + 1.$$

Hence,

$$\prod_{j=1}^{r}(f_j - h) \in \mathcal{L}\left(rG - \sum_{i=1}^{n} P_i\right) \subseteq \mathcal{L}\left(cG - \sum_{i=1}^{n} P_i\right) = \{0\}.$$

It follows that $\prod_{j=1}^{r}(f_j - h) = 0$, and so $h = f_j$ for some j with $1 \leq j \leq r$. Therefore, $\psi(h) = \psi(f_j) \in A$. $\square$

We see from Proposition 6.5.10 that it is crucial to find a positive divisor G of F with $\deg(G) < n$ and $\mathcal{L}(cG - \sum_{i=1}^{n} P_i) = \{0\}$. Lemma 6.5.12 below provides a sufficient condition for the existence of such a divisor G. First we need another auxiliary result. We recall the notation $\operatorname{Cl}(F)$ for the group of divisor classes of degree 0 of F, $h(F) = |\operatorname{Cl}(F)|$ for the divisor class number of F, and $\mathcal{A}_m$ for the set of positive divisors of F of degree m.

Lemma 6.5.11. Let F be a global function field of genus g with a rational place P_0. Then for any fixed integers $c \geq 1$ and $m \geq g$, the subset of $\mathrm{Cl}(F)$ given by

$$\{c[H - \deg(H)P_0] : H \in \mathcal{A}_m\}$$

has cardinality at least $h(F)/c^{2g}$.

Proof. Consider first any divisor class $[D]$ in $\mathrm{Cl}(F)$. Then $\deg(D) = 0$ and Riemann's theorem yields

$$\ell(D + m P_0) \geq m + 1 - g \geq 1.$$

Thus, we can choose a nonzero $f \in \mathcal{L}(D + m P_0)$. Then

$$G := \mathrm{div}(f) + D + m P_0 \geq 0,$$

and hence $G \in \mathcal{A}_m$. Furthermore, D is equivalent to $G - m P_0$, and so $[D] = [G - \deg(G)P_0]$. Thus, we have shown that

$$\{[H - \deg(H)P_0] : H \in \mathcal{A}_m\} = \mathrm{Cl}(F).$$

Now consider the group endomorphism γ of $\mathrm{Cl}(F)$ given by $\gamma([D]) = c[D]$ for all $[D] \in \mathrm{Cl}(F)$. We use the fact that, for any prime p, the p-rank of $\mathrm{Cl}(F)$ is at most $2g$ (see [83, p. 39] and [107, Chapter 11]). Note that the p-rank of $\mathrm{Cl}(F)$ is the number of summands that occur in a direct decomposition of the p-Sylow subgroup of $\mathrm{Cl}(F)$ into cyclic components. It follows then by using the structure theorem for finite abelian groups that the kernel of γ has order at most c^{2g}, and so the result of the lemma is established. $\qquad\square$

We recall that $A_m(F)$ denotes the number of positive divisors of F of degree m. In other words, we have $A_m(F) = |\mathcal{A}_m|$.

Lemma 6.5.12. Let F be a global function field of genus g with $N(F) \geq 1$. Let c, m, and n be integers with $c \geq 1$, $m \geq g$, and $cm \geq n$. Assume that $A_{cm-n}(F) < h(F)/c^{2g}$. Then, given a divisor D of F with $\deg(D) = n$, there exists a positive divisor G of F with $\deg(G) = m$ such that $\mathcal{L}(cG - D) = \{0\}$.

Proof. Choose a rational place P_0 of F and consider the two subsets of $\mathrm{Cl}(F)$ defined by

$$T_1 = \{[cH - c\deg(H)P_0] : H \in \mathcal{A}_m\},$$
$$T_2 = \{[K + D - \deg(K + D)P_0] : K \in \mathcal{A}_{cm-n}\}.$$

Then we have

$$|T_2| \le A_{cm-n}(F) < \frac{h(F)}{c^{2g}} \le |T_1|,$$

where we used Lemma 6.5.11 in the last step. Thus, there exists a divisor class $[cG - c\deg(G)P_0]$ with $G \in \mathcal{A}_m$, which does not belong to T_2. We claim that $\mathcal{L}(cG - D) = \{0\}$. Otherwise, there would be a nonzero $f \in \mathcal{L}(cG - D)$. Then

$$K := \mathrm{div}(f) + cG - D \ge 0$$

and $\deg(K) = cm - n$. Note that cG is equivalent to $K + D$, and so

$$[cG - c\deg(G)P_0] = [K + D - \deg(K + D)P_0].$$

This contradicts the choice of G. $\square$

After these preparations, we are now in a position to prove a lower bound on $D_q(c)$ due to Xing [132], which improves on Theorem 6.5.9 in a certain range for the parameter c.

Theorem 6.5.13. For any prime power $q > 4$ and any integer c with $2 \le c < \sqrt{q}$ we have

$$D_q(c) \ge \frac{1}{c} - \frac{1}{A(q)} + \frac{1 - 2\log_q c}{cA(q)}.$$

Proof. We can assume that

$$A(q) - c + 1 - 2\log_q c > 0, \tag{6.1}$$

for otherwise the result is trivial. Let $F_1/\mathbb{F}_q$, $F_2/\mathbb{F}_q$, ... be a sequence of global function fields such that $g_i := g(F_i)$ and $n_i := N(F_i)$ satisfy

$$\lim_{i \to \infty} g_i = \infty \qquad \text{and} \qquad \lim_{i \to \infty} \frac{n_i}{g_i} = A(q).$$

Let $\mathcal{P}_i$ be the set of all rational places of F_i and put

$$D_i = \sum_{P \in \mathcal{P}_i} P.$$

We note the following consequence of Lemma 5.3.4 with $s = 0$: there exists an absolute constant $A > 0$ such that for any global function field $F/\mathbb{F}_q$ of genus $g \geq 1$ and any integer $b \geq 0$ the number $A_b(F)$ of positive divisors of F of degree b satisfies

$$A_b(F) < Ah(F)gq^{b-g}. \tag{6.2}$$

Now choose a real number ε with

$$0 < \varepsilon < \min(1 - 2\log_q c, \ A(q) - c + 1 - 2\log_q c).$$

This choice is possible by (6.1) and the condition on c. For $i = 1, 2, \ldots$ put

$$m_i = \left\lfloor \frac{n_i + (1 - 2\log_q c - \varepsilon)g_i}{c} \right\rfloor.$$

Then for sufficiently large i we have

$$0 \leq cm_i - n_i \leq (1 - 2\log_q c)g_i - \log_q(Ag_i).$$

It follows then from (6.2) that

$$A_{cm_i - n_i}(F_i) < Ah(F_i)g_i q^{cm_i - n_i - g_i} \leq \frac{h(F_i)}{c^{2g_i}}.$$

for sufficiently large i. Furthermore, from $\varepsilon < A(q) - c + 1 - 2\log_q c$ we obtain

$$\lim_{i\to\infty} \frac{m_i}{g_i} = \frac{A(q) + 1 - 2\log_q c - \varepsilon}{c} > 1,$$

and so $m_i \geq g_i$ for sufficiently large i. Therefore, Lemma 6.5.12 shows that for sufficiently large i there exists a positive divisor G_i of F_i with $\deg(G_i) = m_i$ and $\mathcal{L}(cG_i - D_i) = \{0\}$. Next we note that

$$\lim_{i\to\infty} \frac{m_i}{n_i} = \frac{1}{c} + \frac{1 - 2\log_q c - \varepsilon}{cA(q)} < \frac{2}{c} \leq 1$$

since $A(q) > c - 1 \geq 1$ by (6.1). Thus, we get $m_i < n_i$ for sufficiently large i. Hence, for sufficiently large i, Proposition 6.5.10 yields the c-frameproof code $C(D_i, G_i)$ over $\mathbb{F}_q$ of length n_i and cardinality $q^{\ell(G_i)} \geq q^{m_i - g_i + 1}$. This implies

$$D_q(c) \geq \lim_{i\to\infty} \frac{m_i - g_i + 1}{n_i} = \frac{1}{c} + \frac{1 - 2\log_q c - \varepsilon}{cA(q)} - \frac{1}{A(q)}.$$

Letting ε tend to 0, we arrive at the desired result.   □

An issue that is somewhat related to frameproof codes is that of secret-sharing schemes. A *secret-sharing scheme* protects confidential data (the "secret," often a password or a cryptographic key) by distributing partial information about the secret to the users in such a way that only certain legitimate coalitions of users can reconstruct the secret. An interesting application of algebraic curves over finite fields to the construction of secret-sharing schemes was recently described by Chen and Cramer [17]. Their secret-sharing scheme generalizes the classical Shamir scheme (see [121, Section 15.2]) which corresponds to the case of genus 0, that is, when the algebraic curve is the projective line. The paper [17] contains also applications to secure protocols for multiparty computations.

6.6 Fast Arithmetic in Finite Fields

The efficient implementation of cryptographic schemes based on finite fields or algebraic curves over finite fields requires fast arithmetic in finite fields. Since the case of small finite fields is trivial, the emphasis in the study of fast finite-field arithmetic is on large finite fields. If $q = p^e$ with a prime p and an integer $e \geq 1$, then $\mathbb{F}_q$ can be represented as the residue class field $\mathbb{F}_p[x]/(P(x))$ with an irreducible polynomial $P(x)$ over $\mathbb{F}_p$ of degree e. Hence, in this representation, addition in $\mathbb{F}_q$ is equivalent to addition of polynomials over $\mathbb{F}_p$, which we consider to be easy from the viewpoint of computational complexity. The interesting operation from this viewpoint is therefore multiplication in large finite fields.

We adopt the computational model of bilinear complexity. Let $\mathbb{F}_q$ be an arbitrary finite field and let $\mathbb{F}_{q^n}$ be its extension field of degree n. We consider multiplication in $\mathbb{F}_{q^n}$. Let $\{\alpha_1, \ldots, \alpha_n\}$ be an ordered basis of $\mathbb{F}_{q^n}$ over $\mathbb{F}_q$. Then for any $\beta, \gamma \in \mathbb{F}_{q^n}$, we have

$$\beta = \sum_{i=1}^{n} b_i \alpha_i, \qquad \gamma = \sum_{i=1}^{n} c_i \alpha_i,$$

where $b_i, c_i \in \mathbb{F}_q$ for $1 \leq i \leq n$. Furthermore, we can write

$$\beta\gamma = \sum_{i=1}^{n} d_i \alpha_i \qquad (6.3)$$

with $d_i \in \mathbb{F}_q$ for $1 \leq i \leq n$. One way of achieving efficient multiplication in $\mathbb{F}_{q^n}$ is to express the d_i in a convenient form in terms of $b_1, \ldots, b_n, c_1, \ldots, c_n$.

Definition 6.6.1. Let q be an arbitrary prime power and let $n \geq 1$ be an integer. The *bilinear complexity* $\mu_q(n)$ is the least positive integer μ for which there exist $a_{ij} \in \mathbb{F}_q$, $1 \leq i \leq n$, $1 \leq j \leq \mu$, and linear forms $L_j^{(1)}(x_1, \ldots, x_n)$ and $L_j^{(2)}(x_1, \ldots, x_n)$ over $\mathbb{F}_q$ for $1 \leq j \leq \mu$ such that for any $\beta, \gamma \in \mathbb{F}_{q^n}$, with the notation above, we can write

$$d_i = \sum_{j=1}^{\mu} a_{ij} L_j^{(1)}(b_1, \ldots, b_n) L_j^{(2)}(c_1, \ldots, c_n) \quad \text{for } 1 \leq i \leq n.$$

It is obvious that $\mu_q(n) \leq n^2$. Chudnovsky and Chudnovsky [19] described a method of obtaining better upper bounds on $\mu_q(n)$ by using global function fields over $\mathbb{F}_q$. Since the case $n = 1$ is trivial, we can assume that $n \geq 2$. Let $F/\mathbb{F}_q$ be a global function field and let Q be a place of F of degree n. Then the residue class field F_Q of Q is isomorphic to $\mathbb{F}_{q^n}$, and so we can study $\mu_q(n)$ by considering the bilinear complexity of multiplication in F_Q.

Proposition 6.6.2. Let $F/\mathbb{F}_q$ be a global function field and let $N \geq 1$ and $n \geq 2$ be integers. Suppose that $N(F) \geq N$ and that $\mathcal{P} = \{P_1, \ldots, P_N\}$ is a set of N distinct rational places of F. Moreover, assume that there exists a place Q of F of degree n. Let D be a divisor of F with $\operatorname{supp}(D) \cap \mathcal{P} = \emptyset$ and $Q \notin \operatorname{supp}(D)$. Furthermore, assume that D satisfies the following two conditions:

(i) the map $\varphi : \mathcal{L}(D) \to F_Q$ given by $\varphi(f) = f(Q)$ for all $f \in \mathcal{L}(D)$ is surjective;

(ii) the map $\psi : \mathcal{L}(2D) \to \mathbb{F}_q^N$ given by

$$\psi(f) = \big(f(P_1), \ldots, f(P_N)\big) \quad \text{for all } f \in \mathcal{L}(2D)$$

is injective.

Then we get

$$\mu_q(n) \leq \ell(2D) \leq N.$$

Proof. Since φ is surjective, we can choose an $\mathbb{F}_q$-basis $\{f_1(Q), \ldots, f_n(Q)\}$ of F_Q by using elements $f_1, \ldots, f_n \in \mathcal{L}(D)$. Note that any product $f_i f_j$, $1 \leq i, j \leq n$, lies in $\mathcal{L}(2D)$. Thus, if we put $t = \ell(2D)$ and let $\{g_1, \ldots, g_t\}$ be an $\mathbb{F}_q$-basis of $\mathcal{L}(2D)$, then we can write

$$f_i f_j = \sum_{k=1}^{t} e_{ijk} g_k \quad \text{for } 1 \leq i, j \leq n$$

with all $e_{ijk} \in \mathbb{F}_q$. Furthermore, we have

$$g_k(Q) = \sum_{i=1}^{n} h_{ik} f_i(Q) \quad \text{for } 1 \leq k \leq t$$

with all $h_{ik} \in \mathbb{F}_q$. Define the bilinear forms

$$B_k(x_1, \ldots, x_n, y_1, \ldots, y_n) = \sum_{i,j=1}^{n} e_{ijk} x_i y_j \quad \text{for } 1 \leq k \leq t.$$

Now we consider the product of two elements $x, y \in F_Q$. We can write

$$x = \sum_{i=1}^{n} b_i f_i(Q), \qquad y = \sum_{i=1}^{n} c_i f_i(Q),$$

where $b_i, c_i \in \mathbb{F}_q$ for $1 \leq i \leq n$. Then

$$xy = \sum_{i,j=1}^{n} b_i c_j (f_i f_j)(Q) = \sum_{i,j=1}^{n} b_i c_j \sum_{k=1}^{t} e_{ijk} g_k(Q)$$

$$= \sum_{k=1}^{t} B_k(b_1, \ldots, b_n, c_1, \ldots, c_n) g_k(Q)$$

$$= \sum_{k=1}^{t} B_k(b_1, \ldots, b_n, c_1, \ldots, c_n) \sum_{i=1}^{n} h_{ik} f_i(Q)$$

$$= \sum_{i=1}^{n} \left(\sum_{k=1}^{t} h_{ik} B_k(b_1, \ldots, b_n, c_1, \ldots, c_n) \right) f_i(Q).$$

This means that if, in analogy with (6.3), d_i denotes the coefficient of $f_i(Q)$ in the $\mathbb{F}_q$-basis representation of xy, then

$$d_i = \sum_{k=1}^{t} h_{ik} B_k(b_1, \ldots, b_n, c_1, \ldots, c_n) \quad \text{for } 1 \leq i \leq n. \qquad (6.4)$$

Note that $t = \ell(2D) \leq N$ by condition (ii) in the proposition. Consider the $N \times t$ matrix

$$A = \left(g_k(P_r) \right)_{1 \leq r \leq N, 1 \leq k \leq t}$$

over $\mathbb{F}_q$. If A had rank less than t, then its columns would be linearly dependent over $\mathbb{F}_q$, and so there would exist $w_1, \ldots, w_t \in \mathbb{F}_q$ not all 0 such that

$$\sum_{k=1}^{t} w_k g_k(P_r) = 0 \quad \text{for } 1 \leq r \leq N.$$

Then for $f = \sum_{k=1}^{t} w_k g_k \in \mathcal{L}(2D)$ we have $f \neq 0$, but $\psi(f) = (0, \ldots, 0) \in \mathbb{F}_q^N$, a contradiction to ψ being injective. Thus, A has rank t and there exists a nonsingular $t \times t$ submatrix A_0 of A with rows corresponding to the distinct rational places $Q_1, \ldots, Q_t \in \mathcal{P}$, that is,

$$A_0 = \left(g_k(Q_r) \right)_{1 \leq r \leq t, 1 \leq k \leq t}.$$

Now put

$$S = A_0^{-1} = (s_{kr})_{1 \leq k \leq t, 1 \leq r \leq t}$$

with all $s_{kr} \in \mathbb{F}_q$. We define the linear forms

$$L_r(x_1, \ldots, x_n) = \sum_{i=1}^{n} f_i(Q_r)x_i \quad \text{for } 1 \leq r \leq t.$$

Then for $1 \leq k \leq t$,

$$\sum_{r=1}^{t} s_{kr} L_r(x_1, \ldots, x_n) L_r(y_1, \ldots, y_n) = \sum_{r=1}^{t} s_{kr} \sum_{i,j=1}^{n} (f_i f_j)(Q_r) x_i y_j$$

$$= \sum_{r=1}^{t} s_{kr} \sum_{i,j=1}^{n} \left(\sum_{u=1}^{t} e_{iju} g_u(Q_r) \right) x_i y_j$$

$$= \sum_{i,j=1}^{n} \left(\sum_{u=1}^{t} e_{iju} \sum_{r=1}^{t} s_{kr} g_u(Q_r) \right) x_i y_j.$$

Using the fact that SA_0 is the identity matrix, we obtain

$$\sum_{r=1}^{t} s_{kr} L_r(x_1, \ldots, x_n) L_r(y_1, \ldots, y_n) = \sum_{i,j=1}^{n} e_{ijk} x_i y_j$$

$$= B_k(x_1, \ldots, x_n, y_1, \ldots, y_n)$$

for $1 \le k \le t$. Therefore, from (6.4) for $1 \le i \le n$,

$$d_i = \sum_{k=1}^{t} h_{ik} \sum_{r=1}^{t} s_{kr} L_r(b_1, \ldots, b_n) L_r(c_1, \ldots, c_n)$$

$$= \sum_{r=1}^{t} \left(\sum_{k=1}^{t} h_{ik} s_{kr} \right) L_r(b_1, \ldots, b_n) L_r(c_1, \ldots, c_n).$$

Definition 6.6.1 implies now that $\mu_q(n) \le t = \ell(2D)$, and the inequality $t \le N$ was already noted earlier in this proof. $\square$

A major issue in the application of Proposition 6.6.2 is the construction of a suitable divisor D of F. To this end, we first establish the following auxiliary result.

Lemma 6.6.3. Let F be a global function field of genus g with $N(F) \ge g + 1$. Then there exists a divisor E of F with $\deg(E) = g - 1$ and $\ell(E) = 0$.

Proof. We first prove that there exists a positive divisor G of F with $\deg(G) = g$ and $\ell(G) = 1$. This is trivial for $g = 0$, so we can assume that $g \ge 1$. Let $P_1, \ldots, P_g$ be g distinct rational places of F. We claim that $\ell(P_j) = 1$ for some j with $1 \le j \le g$. Otherwise, we would have $\ell(P_i) > 1$ for $1 \le i \le g$, and so for each of these i we could choose $f_i \in \mathcal{L}(P_i) \setminus \mathbb{F}_q$. Since $v_{P_i}(f_i) = -1$ and $v_{P_k}(f_i) \ge 0$ for $1 \le k \le g$ with $k \ne i$, the triangle inequality implies that the $g + 1$ elements $1, f_1, \ldots, f_g$ are linearly independent over $\mathbb{F}_q$. Choose a divisor $H \ge P_1 + \cdots + P_g$ with $\deg(H) = 2g - 1$. Then $1, f_1, \ldots, f_g \in \mathcal{L}(H)$, hence $\ell(H) \ge g + 1$. On the other hand, $\ell(H) = \deg(H) + 1 - g = g$ by the Riemann-Roch theorem, and we have a contradiction. Thus, the existence of a place P_j with $\ell(P_j) = 1$ is shown. Given $j = j_1$ with $\ell(P_{j_1}) = 1$, we can use a similar argument to obtain a j_2 with $1 \le j_2 \le g$ such that $\ell(P_{j_1} + P_{j_2}) = 1$. This procedure of adding another rational place can be continued as long as

$$\deg(P_{j_1} + \cdots + P_{j_r}) \le g - 1,$$

that is, as long as $r \leq g - 1$. In the last step we then obtain a positive divisor

$$G = P_{j_1} + \cdots + P_{j_g}$$

with $\deg(G) = g$ and $\ell(G) = 1$.

Since there are at most g places of F in the support of G, there exists a rational place P of F with $P \notin \text{supp}(G)$. If we put $E = G - P$, then $\deg(E) = g - 1$. Furthermore, for $f \in \mathcal{L}(E)$ with $f \neq 0$ we would have $f \in \mathcal{L}(G) = \mathbb{F}_q$ and $v_P(f) \geq 1$, a contradiction. Therefore $\ell(E) = 0$.

$\square$

We can now prove the following result obtained by Ballet [2] on the basis of Proposition 6.6.2.

Theorem 6.6.4. Let q be a prime power and let $n \geq 2$ be an integer. Suppose that there exists a global function field $F/\mathbb{F}_q$, which has a place of degree n and satisfies $N(F) \geq 2n + 2g - 1$, where g is the genus of F. Then

$$\mu_q(n) \leq 2n + g - 1.$$

Proof. Note that $N(F) \geq 2n + 2g - 1 > 2g + 1 \geq g + 1$. Thus, we can apply Lemma 6.6.3 and obtain a divisor E of F with $\deg(E) = g - 1$ and $\ell(E) = 0$. Let Q be a place of F of degree n and put $C = E + Q$. Let $\mathcal{R} = \{R_1, \ldots, R_{N(F)}\}$ be the set of all rational places of F. By the approximation theorem (see Theorem 1.5.18), there exists $u \in F$ with $v_{R_i}(u) = v_{R_i}(C)$ for $1 \leq i \leq N(F)$ and $v_Q(u) = v_Q(C)$. Then for $D = C - \text{div}(u)$ we have $\text{supp}(D) \cap \mathcal{R} = \varnothing$ and $Q \notin \text{supp}(D)$. We note also that $\deg(D) = n + g - 1$.

Let $\psi' : \mathcal{L}(2D) \to \mathbb{F}_q^{N(F)}$ be the map defined by

$$\psi'(f) = (f(R_1), \ldots, f(R_{N(F)})) \quad \text{for all } f \in \mathcal{L}(2D).$$

If $f \in \mathcal{L}(2D)$ with $\psi'(f) = (0, \ldots, 0) \in \mathbb{F}_q^{N(F)}$, then $v_{R_i}(f) \geq 1$ for $1 \leq i \leq N(F)$. It follows that $f \in \mathcal{L}(2D - \sum_{i=1}^{N(F)} R_i)$. But

$$\deg\left(2D - \sum_{i=1}^{N(F)} R_i\right) = 2n + 2g - 2 - N(F) < 0,$$

and so $f = 0$ by Corollary 3.4.4. Hence ψ' is injective. Note that $\deg(2D) = 2n + 2g - 2 > 2g - 1$, and so $\ell(2D) = \deg(2D) + 1 - g = 2n + g - 1 =: N$ by the Riemann-Roch theorem. If $\{h_1, \ldots, h_N\}$ is an $\mathbb{F}_q$-basis of $\mathcal{L}(2D)$, then $\{\psi'(h_1), \ldots, \psi'(h_N)\}$ is an $\mathbb{F}_q$-basis of the image of ψ'. It follows that the $N \times N(F)$ matrix

$$B = (h_r(R_i))_{1 \le r \le N, 1 \le i \le N(F)}$$

over $\mathbb{F}_q$ has rank N. Consequently, there exists a set $\mathcal{P} = \{P_1, \ldots, P_N\}$ of N distinct rational places of F such that the columns of B corresponding to the places in $\mathcal{P}$ are linearly independent over $\mathbb{F}_q$. Therefore, the map $\psi : \mathcal{L}(2D) \to \mathbb{F}_q^N$ in condition (ii) of Proposition 6.6.2 is injective.

Next we consider the map $\varphi : \mathcal{L}(D) \to F_Q$ in condition (i) of Proposition 6.6.2. If $f \in \mathcal{L}(D)$ with $\varphi(f) = 0$, then $v_Q(f) \ge 1$, and so $f \in \mathcal{L}(D - Q)$. But $\ell(D - Q) = \ell(E) = 0$, hence, $f = 0$ and φ is injective. This implies $\ell(D) \le \deg(Q) = n$. On the other hand, $\ell(D) \ge \deg(D) + 1 - g = n$ by Riemann's theorem, and so $\ell(D) = n$. Hence, φ is surjective.

Thus, all hypotheses in Proposition 6.6.2 are satisfied and this proposition yields $\mu_q(n) \le N = 2n + g - 1$. $\square$

Corollary 6.6.5. Let $n \ge 1$ be an integer and let q be a prime power with $q \ge 2n - 2$. Then $\mu_q(n) = 2n - 1$.

Proof. The case $n = 1$ is trivial. For $n \ge 2$ we apply Theorem 6.6.4 with F being the rational function field over $\mathbb{F}_q$. Note that $N(F) = q + 1 \ge 2n - 1$, and so all conditions of Theorem 6.6.4 are fulfilled. Hence, we get $\mu_q(n) \le 2n - 1$. Since we always have $\mu_q(n) \ge 2n - 1$ (see [14, Chapter 17]), we obtain indeed $\mu_q(n) = 2n - 1$ whenever $q \ge 2n - 2$. $\square$

According to a well-known result (see [117, Corollary V.2.10]), a global function field $F/\mathbb{F}_q$ of genus g has a place of given degree n whenever

$$2g + 1 \le q^{(n-1)/2}(q^{1/2} - 1).$$

This condition can therefore replace the assumption in Theorem 6.6.4 that $F/\mathbb{F}_q$ has a place of degree n. Further results in Ballet [2] show that if q is a square ≥ 9, then the hypotheses in Theorem 6.6.4 can be satisfied for all sufficiently large n. For more recent work on $\mu_q(n)$, we refer to Ballet [3] and Ballet and Chaumine [4].

Appendix

A.1 TOPOLOGICAL SPACES

For the convenience of the reader, we collect some definitions and facts from topology that are used in this book.

Definition A.1.1. A *topology* on a nonempty set X is a collection $\mathcal{A}$ of subsets of X, called the *open sets* of the topology, which satisfies the following properties:

(i) X and the empty set are open (namely they belong to $\mathcal{A}$);
(ii) the union of any collection of open sets from $\mathcal{A}$ is again an open set;
(iii) the intersection $U \cap V$ of two open sets $U, V \in \mathcal{A}$ is open.

A *topological space* $(X; \mathcal{A})$ is a nonempty set X together with a topology $\mathcal{A}$ on X.

By Definition A.1.1(iii), the intersection of finitely many open sets is again open. On any nonempty set X there are two topologies that come for free: the trivial topology in which the only open sets are $\varnothing$ and X and the discrete topology in which all subsets of X are open. We will often write X for a topological space $(X; \mathcal{A})$.

Definition A.1.2. The complements of open sets are called *closed sets*, that is, $X \setminus U$ is a closed set for any open subset U of a topological space X.

Remark A.1.3. Let X be a topological space and let $\mathcal{B}$ be the collection of its closed sets. Then it is easy to show the following:

(i) $\varnothing$ and X belong to $\mathcal{B}$, that is, they are closed sets;
(ii) an arbitrary intersection of sets from $\mathcal{B}$ belongs to $\mathcal{B}$;
(iii) a finite union of sets from $\mathcal{B}$ is in $\mathcal{B}$.

Conversely, a collection $\mathcal{B}$ of subsets of a nonempty set X satisfying the above three conditions defines a topology $\mathcal{A} = \{X \setminus V\}_{V \in \mathcal{B}}$. In verifying these statements, one uses the well-known properties from set theory that $X \setminus \cap_i U_i = \cup_i (X \setminus U_i)$ and $X \setminus \cup_i U_i = \cap_i (X \setminus U_i)$.

Definition A.1.4. Let X be a topological space and let Y be a subset of X. The *closure* of Y, denoted by $\overline{Y}$, is defined to be the intersection $\cap_{i \in I} B_i$, where $\{B_i\}_{i \in I}$ is the collection of all closed sets containing Y. It is obvious that $\overline{Y}$ is the smallest closed set containing Y. The subset Y is said to be *dense* in X if $\overline{Y} = X$.

Definition A.1.5. Let $(X; \mathcal{A})$ be a topological space and let Y be a nonempty subset of X. Then we have an *induced topology* $(Y; \mathcal{A}|_Y)$ on Y, where $\mathcal{A}|_Y = \{U \cap Y : U \in \mathcal{A}\}$.

Indeed, it is easy to verify that the induced topology $(Y; \mathcal{A}|_Y)$ is a topological space.

Definition A.1.6. Let X and Y be two topological spaces and let f be a map from X to Y. We say that f is *continuous* if $f^{-1}(U)$ is open for any open subset U of Y.

Remark A.1.7. Let X and Y be two topological spaces and let f be a map from X to Y. Then one can easily show that f is continuous if and only if $f^{-1}(V)$ is closed for any closed subset V of Y.

Definition A.1.8. Let S be a nonempty subset of a topological space X with the induced topology. Then S is called *irreducible* if it cannot be expressed as the union $S = S_1 \cup S_2$ of two proper closed subsets S_1 and S_2 of S in the induced topology.

Theorem A.1.9.

(i) Any nonempty open subset of an irreducible topological space is dense and irreducible.

(ii) If S is an irreducible subset of a topological space X, then its closure $\overline{S}$ in X is also irreducible.

(iii) Let S be an open subset of an irreducible topological space X. Then for an irreducible subset T of X, the intersection $S \cap T$ is either empty or irreducible.

(iv) The intersection of any two nonempty open subsets of an irreducible topological space is nonempty.

Proof.

(i) Let X be an irreducible topological space and let U be a nonempty open subset of X. Then $X = \overline{U} \cup (X \setminus U)$. Since $X \setminus U \neq X$, we must have $\overline{U} = X$, that is, U is dense.

   Write $U = U_1 \cup U_2$ for two closed subsets U_1 and U_2 of U in the induced topology. Then there exist two closed subsets V_1 and V_2 of X such that $V_i \cap U = U_i$ for $i = 1, 2$. Hence, $U \subseteq V_1 \cup V_2$. This yields $X = \overline{U} \subseteq V_1 \cup V_2$, and so $X = V_1 \cup V_2$. We conclude that $V_1 = X$ or $V_2 = X$; therefore, $U_1 = U$ or $U_2 = U$.

(ii) Write $\overline{S} = S_1 \cup S_2$ for two closed subsets S_1 and S_2. Then

$$S = S \cap \overline{S} = (S \cap S_1) \cup (S \cap S_2).$$

Hence, $S \cap S_1 = S$ or $S \cap S_2 = S$, that is, $S \subseteq S_1$ or $S \subseteq S_2$. Therefore, $\overline{S} \subseteq S_1$ or $\overline{S} \subseteq S_2$. This means that $\overline{S}$ is irreducible.

(iii) Note that $S \cap T$ is an open subset of the irreducible topological space T. Thus, if $S \cap T$ is nonempty, then it is irreducible by (i).

(iv) Let X be an irreducible topological space and let U_1 and U_2 be two nonempty open subsets of X. If $U_1 \cap U_2$ were empty, then $(X \setminus U_1) \cup (X \setminus U_2) = X$, a contradiction to X being irreducible.

$\square$

Definition A.1.10. If X is a topological space, then the *dimension* of X, denoted by $\dim(X)$, is defined to be the supremum of all nonnegative integers n such that there exists a chain $S_0 \subsetneq S_1 \subsetneq \cdots \subsetneq S_n$ of distinct irreducible closed subsets of X.

Theorem A.1.11.

(i) If Y is a nonempty subset of a topological space X, then $\dim(Y) \leq \dim(X)$.

(ii) If a topological space X is covered by a family $\{U_i\}$ of nonempty open subsets, then $\dim(X) = \sup_i \dim(U_i)$.

(iii) Let Y be a nonempty closed subset of an irreducible finite-dimensional topological space X. If $\dim(X) = \dim(Y)$, then $X = Y$.

Proof.

(i) Let $S_0 \subsetneq S_1 \subsetneq \cdots \subsetneq S_n$ be a chain of distinct irreducible closed subsets of Y. Then we get a chain $\overline{S_0} \subseteq \overline{S_1} \subseteq \cdots \subseteq \overline{S_n}$ of closed subsets of X, where $\overline{S_i}$ is the closure of S_i in X. By Theorem A.1.9(ii), every $\overline{S_i}$ is irreducible. It is clear that $\overline{S_i} \neq \overline{S_{i+1}}$, for otherwise $S_i = Y \cap \overline{S_i} = Y \cap \overline{S_{i+1}} = S_{i+1}$.

(ii) By (i) we have $\dim(X) \geq \sup_i \dim(U_i)$. Let $S_0 \subsetneq S_1 \subsetneq \cdots \subsetneq S_n$ be a chain of distinct irreducible closed subsets of X. Then there exists an open subset U_i such that $S_0 \cap U_i$ is nonempty. Thus, by Theorem A.1.9(iii) we get a chain $(S_0 \cap U_i) \subseteq (S_1 \cap U_i) \subseteq \cdots \subseteq (S_n \cap U_i)$ of irreducible closed subsets of U_i. It suffices now to show that $S_j \cap U_i \neq S_{j+1} \cap U_i$ for $j = 0, 1, \ldots, n - 1$. Suppose that $S_j \cap U_i = S_{j+1} \cap U_i$ for some j. Then we have

$$\begin{aligned} S_{j+1} &= (S_{j+1} \cap U_i) \cup (S_{j+1} \cap (X \setminus U_i)) \\ &= (S_j \cap U_i) \cup (S_{j+1} \cap (X \setminus U_i)) \\ &\subseteq S_j \cup (S_{j+1} \cap (X \setminus U_i)) \subseteq S_{j+1}. \end{aligned}$$

Hence, $S_{j+1} = S_j \cup (S_{j+1} \cap (X \setminus U_i))$. Since S_{j+1} is irreducible, we must have $S_{j+1} \cap (X \setminus U_i) = S_{j+1}$ as $S_j \neq S_{j+1}$. This forces $S_{j+1} \cap U_i = \varnothing$. This is a contradiction since $\varnothing \neq S_0 \cap U_i \subseteq S_{j+1} \cap U_i$.

(iii) Let $n = \dim(Y)$ and let $T_0 \subsetneq T_1 \subsetneq \cdots \subsetneq T_n$ be a chain of distinct irreducible closed subsets of Y. If we had $Y \neq X$, then we would get a longer chain $T_0 \subsetneq T_1 \subsetneq \cdots \subsetneq T_n \subsetneq X$ of distinct irreducible closed subsets of X. By definition, $\dim(X) \geq n + 1 = \dim(Y) + 1$. This is a contradiction. $\qquad\square$

A.2 KRULL DIMENSION

Here and in the following section, all rings are assumed to be commutative with identity.

Definition A.2.1. Let A be a ring. The *height* $\mathrm{ht}(\wp)$ of a prime ideal $\wp$ of A is the supremum of all nonnegative integers n such that there exists a chain $\wp_0 \subsetneq \wp_1 \subsetneq \cdots \subsetneq \wp_n = \wp$ of distinct prime ideals of A.

The *Krull dimension* dim(A) of A is defined to be the supremum of the heights of all prime ideals of A.

The reader may refer to Iitaka [60, pp. 22–25] and Matsumura [78, Section 14] for the proof of the following theorem.

Theorem A.2.2. Let k be a field and let A be an integral domain, which is a finitely generated k-algebra. Then:

(i) the Krull dimension of A is equal to the transcendence degree over k of the quotient field of A;

(ii) for any prime ideal $\wp$ of A, we have

$$\mathrm{ht}(\wp) + \dim(A/\wp) = \dim(A).$$

A.3 DISCRETE VALUATION RINGS

Definition A.3.1.

(i) An integral domain R is called a *local ring* if all elements except units form an ideal of R.

(ii) An integral domain R is called a *discrete valuation ring*, abbreviated DVR, if there exists an irreducible element $t \subset R$ such that every nonzero element z of R can be written in the form $z = ut^m$ for some unit $u \in R$ and a nonnegative integer m. The element t is called a *local parameter* of R.

It is easily seen that the representation $z = ut^m$ in Definition A.3.1(ii) is unique. In the following lemma, we collect some basic properties of local rings and DVRs.

Lemma A.3.2.

(i) If R is a local ring, then there is a unique maximal ideal of R and it consists of all nonunits of R.

(ii) If R is a DVR with a local parameter t, then R is a local ring and its maximal ideal is $\{ut^m \cdot u \text{ is a unit of } R, \ m \geq 1\} \sqcup \{0\}$.

(iii) A DVR has Krull dimension 1.

Proof.

(i) Let $\mathfrak{m}$ be the ideal consisting of all nonunits of R. Let I be a proper ideal of R. It is clear that I does not contain any units, hence, $I \subseteq \mathfrak{m}$.

(ii) It is easy to verify that the set $\mathfrak{m} := \{ut^m : u$ is a unit of R, $m \geq 1\} \cup \{0\}$ forms an ideal of R. It is also clear that $\mathfrak{m}$ is exactly the set of all nonunits of R. Hence, R is a local ring and the rest follows from (i).

(iii) Let t be a local parameter of a DVR R. It is easy to see that every nonzero proper ideal of R is of the form $I_k := \{ut^m : u$ is a unit of R, $m \geq k\} \cup \{0\} = t^k R$ for some integer $k \geq 1$. It is then clear that I_1 is the unique nonzero prime ideal of R, and, hence, the height of I_1 is 1 as $\{0\} \subsetneq I_1$. The desired result follows from Definition A.2.1.   □

Proposition A.3.3. Let R be a DVR with quotient field K. If S is also a DVR with $R \subseteq S \subseteq K$, then $R = S$.

Proof. Let t be a local parameter of R. Then it is easy to see that the quotient field K is $\{ut^m : u$ is a unit of R, $m \in \mathbb{Z}\} \cup \{0\}$. Suppose $R \neq S$. Then there exists an element $x \in S \setminus R$. Hence, $x = vt^{-n}$ for some unit v of R and an integer $n \geq 1$. Let s be a local parameter of S. Then $x = ws^m$ for some unit w of S and an integer $m \geq 0$. As $t \in R \subseteq S$, we can write $t = us^\ell$ for some unit u of S and an integer $\ell \geq 1$. Then $x = ws^m = vu^{-n}s^{-\ell n}$, that is, the unit $vw^{-1}u^{-n}$ is equal to the nonunit $s^{m+\ell n}$. This contradiction shows that $R = S$.   □

Let R be a local ring and let $\mathfrak{m}$ be the unique maximal ideal of R. Then the factor ring $R/\mathfrak{m}$ is a field, denoted by k. Thus, $\mathfrak{m}/\mathfrak{m}^2$ can be viewed as a vector space over k. The field k is called the *residue field* of R.

Proposition A.3.4. If R is a Noetherian local ring of Krull dimension 1 with maximal ideal $\mathfrak{m}$ and residue field k, then R is a DVR if and only if $\dim_k(\mathfrak{m}/\mathfrak{m}^2) = 1$.

Proof. Assume that R is a DVR and let t be a local parameter of R. Then it is easy to see that $\mathfrak{m}/\mathfrak{m}^2 = tR/t^2R \simeq R/tR = k$. Hence, $\dim_k(\mathfrak{m}/\mathfrak{m}^2) = 1$.